NEOMANIA

Neomania

How Our Obsession With Innovation is Failing Science, and How to Restore Trust

Krist Vaesen

OpenBook Publishers

https://www.openbookpublishers.com

All external links were active at the time of publication unless otherwise stated and have been archived via the Internet Archive Wayback Machine at https://archive.org/web

Digital material and resources associated with this volume are available at https://doi.org/10.11647/OBP.0507#resources

Information about any revised edition of this work will be provided at https://doi.org/10.11647/OBP.0507

ISBN Paperback: 978-1-80511-781-0
ISBN Hardback: 978-1-80511-782-7
ISBN Digital (PDF): 978-1-80511-783-4
ISBN HTML: 978-1-80511-785-8

DOI: 10.11647/OBP.0507

Cover image: Photo by Hush Naidoo Jade Photography on Unsplash, https://unsplash.com/photos/steel-wool-photography-of-yellow-lights-QK9BcrMeXK4

Cover design: Jeevanjot Kaur Nagpal

My intention is not to prove that hitherto I have been right; but to discover whether I am right [...] *Yes, we will question everything, everything once again. And we shall advance not in seven-league boots, but at a snail's pace. And what we find today we shall strike out from the record tomorrow, and only write it in again when we have once more discovered it. And what we wish to find, if we do find it, we shall regard with especial distrust.*

Galileo in Bertolt Brecht's *Life of Galileo*

Dedications available upon request

Table of Contents

Krist Vaesen is an Associate Professor of Philosophy of Innovation at Eindhoven University of Technology (the Netherlands) and serves as director and co-founder of META/e—the Eindhoven Meta-science Center. This interdisciplinary center focuses on the scientific study of science itself, with expertise in areas such as open science, reproducibility, team science, and the role of AI in research. *Neomania* owes much to the many insightful conversations with members of META/e.

List of illustrations

List of tables

Introduction

A key reason for writing this book is personal more than anything else. When I started working on it, I had just recovered from an episode of depression—I had been out for close to six months.

In this period, I had spent much of my time in bed, wondering why my energy levels were so soberingly low. I couldn't find a cause in my private life and so concluded that the problem had to be job-related. But which aspect of my job was to blame? What had been different in the months or years leading up to my illness?

I hadn't been overburdened with work, nor had I had too little to do—my workload seemed to have been just right. Neither had our research group gone through some kind of storm, organizational or otherwise. No big egos blowing the vibe, no toxic leadership, no major financial worries, nothing—on the contrary, even.

It took me a long while to figure out that one thing in actuality *had* changed: for almost two years, I had done little to no teaching. I generally like interacting with students, so perhaps I was just missing them? Perhaps the instant gratification associated with teaching was what kept me going, what allowed me to carry out my other, more delayed-gratification tasks, such as research and admin?

It took me another long while to realize that, if at all, this could be only a partial explanation. The exhaustion I experienced felt at odds with having done too little of something. I really had done *too much* of something. But what?

About two years before I fell ill, my department had kindly asked me whether I would mind serving as one of its dedicated grant proposal writers. The deal was this: the department would hire a replacement teacher, so that I could spend virtually all of my time on grant acquisition. Because it was implied that I had been selected for my superior grantsmanship, I felt honoured, and gratefully accepted the deal. In

https://doi.org/10.11647/OBP.0507.00

the period that followed, I wrote a secret, but certainly big number of research proposals. Any academic will tell you that grant-writing consumes tons of energy—*voilà*, the explanation for my exhaustion!

...Or not? Sure, gratification is extremely delayed in grant-writing. Not only are success rates criminally low (10% or lower is not uncommon), assessment procedures may take what feels like ages (1.5 years is not uncommon). Getting your project started may take an additional 0.5–1 years, carrying out the proposed plan another 1–5 years. The total time between an academic's Eureka!-moment and the publication of their brilliant idea thus easily adds up to 7 years.

But, academics have a coping strategy: they constantly feed their so-called academic pipeline. The idea is straightforward. If publication is your ultimate form of gratification, and you need gratification once in a while, make sure that, once in a while, you feed the pipeline with a novel manuscript (viz., submit it for review to a journal); if success in grant-writing is your ultimate form of gratification, do the same for grant proposals (viz., regularly submit a proposal for review to a funding agency).

My pipeline was full, with papers and proposals, and often enough it spit out a success. So as a serial grant-writer, I wasn't suffering from delayed gratification any more than my peers—surely not the many who, due to a crazily heavy teaching load, were not in a position to feed the pipeline at all.

Midway through my sick leave, I stumbled across a book that did more than any therapist had done in getting my diagnosis right: *Innovation for the Fatigued: How to Build a Culture of Deep Creativity*, by Alf Rehn. Before you think I like this book, I genuinely don't (see f.n. 2). But the book definitely had one merit: it gave a name to the illness I was suffering from. I was suffering from *innovation fatigue*, a condition that Rehn described as follows:

> In a contemporary organization, the average employee has heard the term [innovation] bandied about more times than they can count, seen numerous innovation consultants pass through, developed a slight allergy to multi-coloured Post-it® notes, and feel [*sic*], on the whole, sick and tired of it all.[1]

---

1 See Rehn, 2019, p. 12.

As a serial grant-writer, this was certainly something I could relate to. Because I had refrained from submitting the same idea to more than one funding agency, I had for two years spawned one novel idea after the other; my institute's internal grant-writing consultants had—and I am genuinely thankful for this—pushed me to the limits of my innovative abilities. I wasn't sick of Post-it® notes but of the countless flow-charts that I had produced to get my new ideas straight and communicate them. Re-inventing yourself is fine every once in a while. Doing so most of the time is not—it leads to alienation and exhaustion.[2]

When returning fully recovered to office, I just had to drop the term 'innovation fatigue', and my colleagues would immediately understand the reason for my sick leave. Not that they knew the term. But they *did* have a pretty good intuitive sense of what it referred to—every one recognized the constant pressure on academics to innovate, and some admitted to having gone through bouts of 'innovation fatigue' themselves.

I wanted to do something with this insight—if only to turn the frustration that my illness had engendered into something more constructive. I considered writing a book about the well-being of academics, about how to prevent them from falling prey to the condition I had been suffering from. But such a book would really go far beyond my expertise as an applied biologist and philosopher of science.

So—as philosophers tend to do when they don't know what else to do—I ran a simple thought experiment. Let's assume that not only I, but all academics, behaved as I did in recent years. That is, let's assume that they all preferred, or were pushed to prefer, the new over the old. What would be the ramifications for science?

Any substantive theory about the growth of scientific knowledge will tell you: it would be a disaster. For ideas to become part of our stock of knowledge, they require time to mature: they need thorough validation, methodological and conceptual refinement, integration with other ideas and theories, and so forth. Neomania—an excessive love for the new—would undermine precisely that which makes scientific knowledge scientific.[3]

2 And the cure is not, as Rehn suggests, to become 'deep' rather than 'shallow' innovators. Only detox works.

3 I borrow the term 'neomania' from Taleb (2012).

Connecting the dots between the results of my, uhm, 'experiment' and something else I had for a while been concerned by then came easily. That other something was the so-called replication crisis. Like many (and many) other researchers, I was seriously taken aback by the evidence that had accumulated over the last decade showing that the results of studies in almost any imaginable scientific discipline are hard, if not impossible, to accurately reproduce. It really is a *minimal* condition for results to be credible: if you run an experiment, and you re-run it a bit later, the results should converge. But apparently, they frequently did not—so much for the credibility of science...

Let's briefly, and for the moment roughly, connect the dots. Suppose that researchers are indeed strongly incentivized to be neomaniacs. How would they behave? For one, they would pursue novel ideas rather than engage in replication work; and that would explain why so little of such work had been carried out until only a decade ago, i.e., when the replication crisis arrived in full swing. For another, because novelty requires priority, researchers would likely adopt strategies to increase their odds at getting out their idea *first*, such as cutting corners and rushing their research. Neomania could then account for the replication crisis in the following way: the original studies could not be reproduced because the original authors had cut corners and rushed their research in an attempt to arrive first.

I had some other deeply disturbing dots lying around: the generalizability crisis, the applicability crisis, the methodology crisis, and the theory crisis—scientific crises which have received far less attention, and which I will detail in later chapters. For now, what matters is that I managed to connect all of them to my first dot: neomania turned out to be a plausible explanation for all of them.

Part I of this book is devoted to telling this story—a story of how our (society's) obsession with novelty and innovation has pushed science into a poly-crisis.

The second part is more cheerful, as I sketch a way out. The path towards a brighter future involves reforms to the way we organize science far more radical than the reforms championed by the open science movement (which I am highly sympathetic to). We should get to a point where science does less, but better. Where scientists coordinate themselves into taking the time to answer the questions that matter most to us. Where collaboration is valued more than competition. Where fewer people suffer from innovation fatigue.

The programme of the book

Chapter 1 establishes that science is in a poly-crisis. I start by reviewing the body of evidence that has led to the replication crisis. Researchers often appear to be unable to obtain results comparable to results obtained by other researchers, even when following the same research procedures. And virtually every scientific discipline seems to be affected: from psychology to biomedicine, from economics to engineering, from sociology to physics. I explain how this undermines the credibility of science: very little remains of the things we thought we knew, especially because the replication crisis is not the only crisis. Indeed, I go on to survey evidence for a methodology crisis, a generalizability crisis, an applicability crisis, and a theory crisis. I conclude that there is something rotten in the state of science.

Chapter 2 puts the blame on our society's neomania. Scientists tend to focus on the study of novel things, and to refrain from activities that increase the robustness of existing knowledge—activities such as replication, methodology checks, generalizability and applicability tests, and theory development. The reason why scientists behave as neomaniacs is that they are prompted to by every salient stakeholder in the scientific enterprise: scientific journals, science funders, universities, businesses, policy makers, politicians, the media, and the general public. As if every wall in every research lab in the world whispers the words 'Thou Shalt Innovate!'.

Chapter 3 provides a history of science's obsession with novelty and innovation. It traces this obsession to the early seventeenth century, which has been aptly called 'The Age of the New'. Science then started increasingly to embrace newness, up to the present point, where (one of) science's principal aim(s) is to serve innovation for the sake of (economic) progress.

Chapter 4, the first chapter of Part II, looks at the open science movement's response to the replication crisis (it has been less concerned with the other crises). Open science may mean a number of things, including, but not limited to, open access publication and openly sharing research protocols and data. The motivation for opening science is twofold. First, it likely increases the quality of published findings; for by sharing, researchers make themselves vulnerable to critique—the mere feeling of being watched plausibly motivates them to do their best to get it right. Second, openness facilitates criticism: it allows others to run additional checks, such as replications, generalizability tests, etc.

I argue that while open science is necessary, it is not enough. In addition to removing the barriers to replication (and other activities that contribute to the process of knowledge maturation), we must also actively entice researchers to engage in replication (and other activities that contribute to the process of knowledge maturation). Data sharing is fine, but pointless without actual reuse.

Chapter 5 introduces scientific coordination as a way of resolving the crises. The current system incentivizes researchers to work on their own isolated islands of novelty, and to move to another island even before the new has become the old. The solution is to gather communities of researchers on a couple of carefully and bottom-up selected islands where they work together, performing all the steps that give us robust knowledge. Put non-metaphorically, science needs to be organized not around research projects but research *programmes*—concerted efforts at addressing challenges that scientists and relevant stakeholders find most pressing.

This implies that scientists need to subordinate their own (neomanic) agenda to the common good—a good that is determined at the community-level in a grassroots manner, and that researchers can voluntarily align themselves with. It also implies that they need to actively and cautiously coordinate their actions, such that the collective is able to run smoothly and work effectively. And, for those who already know the phenomenon, it implies that we need to go beyond big team science (as it is currently practiced).

Chapter 6 then describes a future where science is effectively organized around research programmes. To start, in this future, science funders facilitate bottom-up research programme selection and development through the organization of consensus conferences with researchers and other stakeholders, provide sustained support for the realization of selected programmes, and do so transnationally.

Also for journals the analytic unit would be the research programme.

Program members submit the plans of their programme for review—if accepted, journals promise to publish all of its outputs (including replications and the like). Because commercial publishers may be reluctant to implement such 'registered reports' for programmes, science funders may need to step up, by launching their own open access publication platforms—and perhaps even *requiring* their funded programmes to publish their outputs there.

Universities, finally, would do a wide variety of things, ranging from accepting the use of program-level rather than individual-level performance criteria in hiring and promotion, to selecting candidates based on their contributions and ability to do programme work, and preparing students to take up such work.

Chapter 7 serves as the epilogue. There I discuss two pressing issues that, in light of recent events, warrant special attention: science's adoption of AI, and the growing anti-scientific sentiment in many liberal democracies. And I end with yet another personal confession.

PART I
THE DIAGNOSIS

1. Science in poly-crisis

Suppose I were to tell you that you, like any other human being, can feel the future. That, when presented with digital pictures of two completely opaque curtains, you can *feel* which curtain has another picture behind it, and which one does not. And that you are particularly good at such pre-cognition when the picture behind the curtain is a pornographic one. Would you believe me that you have this ability?

I guess—or hope—not. But that's precisely the thing that a 2011 paper, published in the reputable *Journal of Personality and Social Psychology*, wanted to make us believe. In the article, entitled 'Feeling the future: Experimental evidence for anomalous retroactive influences on cognition and affect', the eminent Cornell psychologist Daryl Bem reported on a series of nine experiments that support the existence of paranormal abilities—like the pornography pre-cognition ability just described.[1] In these studies, participants engaged in traditional experimental psychology tasks with a unique twist: rather than responding to stimuli that had already been presented, participants needed to respond to a stimulus that would be presented in the future. For example, instead of indicating whether a picture appeared on the left or right side of the screen, participants had to indicate where a picture would appear in the future. Performance statistically better than guessing was taken to indicate that people had the ability to predict the future. This claim contradicts not only common sense, but also all working models of physics. And yet, Bem presented a series of studies, including over 1,000 subjects and deploying widely used psychological protocols, where subjects performed better than guessing (i.e., above random probability). Psychologists were quick to question Bem's—not unusual—protocols. Simmons, Nelson, and Simonsohn, for example,

1 See Bem, 2011.

 https://doi.org/10.11647/OBP.0507.01

convincingly explained how these protocols could substantially inflate false positive rates (a 'false positive' refers to erroneously finding a statistically significant effect where, in actuality, there is no such effect).[2]

Suppose, now, I were to tell you that your behaviour is determined by unconscious factors. Particularly, that if I were to expose you in the lab to words related to old age, you would walk out of the lab slower than if you hadn't been exposed to these words. Based on a famous, widely cited study published in 1996 by Bargh, Chen and Burrows,[3] you should believe me. That study found evidence of such unconscious priming. However, in 2012—around the same time that Bem's precognition paper came out—a team of Belgian scientists led by Stéphane Doyen reported that they had not been able to replicate Bargh, Chen and Burrows' results.[4] So, despite their attempt at closely following Bargh et al.'s experimental protocol, the replication study found no priming effect. Or rather—even worse—Doyen et al. *did* find a priming effect, not among the study subjects but instead among the *experimenters*: if the latter had been primed to believe there would be an effect on walking speed, they would observe one in the experiments; but they would not observe this if they hadn't been primed. All this was not much to the liking of Bargh, who attacked Doyen and his colleagues in a series of blog posts, accusing them of methodological and other flaws.[5]

Suppose, finally, that I were to tell you that when you pick up a scientific journal, even a reputable one, and read a random article that reports positive findings, chances are high that, given current statistical practices, the article's results are false. Would you believe me? Again—unfortunately—you should. And why this is so was explained by John Ioannidis in a series of papers published between 2005 and 2014.[6] Ioannidis' argument is quite technical, but here is the gist of it (for the technical details, see Text Box 1.1).

You have probably heard about p-values, and about the fact that scientists often set their statistical significance level at 0.05—such that they reject the null hypothesis if the p-value is lower than 0.05. What this

2 See Simmons et al., 2011.
3 See Bargh et al., 1996. As of 17 January 2024, the paper had been cited over 6,400 times.
4 See Doyen et al., 2012.
5 See Leys, 2024.
6 See Ioannidis, 2005, 2008; Ioannidis et al., 2014.

basically means is that there is only a 5% probability that they will reject the null hypothesis when the null hypothesis is in actuality true. You might think that, accordingly, at least in the long run, 95% of published findings must be true positives. But that is too quick, Ioannidis points out. This percentage will be much lower if one takes into account statistical power (the probability of a significance test detecting an effect when there actually is one), prior probabilities (the fact that some hypotheses are more likely to be true than others), publication bias (i.e., the fact that journals are much more likely to publish positive than negative results),[7] and active manipulation (e.g., removing outlier data). For instance, until 2011 it was common in psychology to publish studies that were under-powered—that is, studies with statistical power as low as 60%, instead of the generally accepted 80%.[8] Supposing, conservatively, that we are investigating hypotheses that are equally likely to be true or false (so, set the prior probability at 0.5), it can be shown that, given a statistical power of 60%, only 92% rather than 95% of our significant positive findings will be true positives (see PPV value in row 1 of the table in Text Box 1.1). This number drops considerably if we are investigating hypotheses that are more implausible (i.e., if we lower the prior probability)—e.g., the hypothesis that people can 'feel' the future. So, if we for instance assume a prior probability of 0.1, only 57% of significant positive results will be true positives (see PPV value in row 2 of the table in Text Box 1.1). Matters get even worse when we take into account publication bias or biases relating to active manipulation. Assuming a 30% bias rate, the total number of true published findings plummets to 19% (see PPV value in row 3 of the table in Text Box 1.1). Of course, not all research fields have low power, study hypotheses with low prior probabilities, and have large bias rates—but Ioannidis' provocative paper was an eye-opener for many researchers that a single significant result provides less certainty than they had previously assumed.

These three research strands—viz., the pre-cognition, the priming, and the Ioannidis strand—led to a sense of crisis among psychological scientists in the early 2010s. Each strand appeared to undermine the credibility of the discipline; it now only seemed a matter of clarifying

7 See Springer Nature, July 2025, https://stories.springernature.com/the-state-of-null-results-white-paper/index.html.

8 See Fraley et al., 2022.

the extent of the crisis. And this was clarified through the efforts of two large research consortia: the Open Science Collaboration and Many Labs.

In 2015 a team of 270 authors published, under the name of the Open Science Collaboration, the results of a research project in which they replicated 100 studies.[9] The original studies had all been published in three high-profile psychology journals in the year 2008. Although the authors of the replication project stayed as close as possible to the original studies, they obtained a replication rate of no more than 36%. So, whilst 97 out of the original 100 studies found a significant effect, this number dropped to 35 out of 100 in the replications. Further, replicated effect sizes—a value measuring the strength of the relationship between two variables—were on average *only half as large* as the effect sizes of the original studies.

The *Reproducibility Project: Psychology* paved the way for other large-scale replication projects in psychology. Researchers working under the Many Labs label did not randomly sample and replicate studies from the scientific literature, but rather directly targeted meta-scientific questions about replicability. For example, *Many Labs 2* explored the impact of diversity in participant characteristics and achieved a replication rate of merely 54%.[10] *Many Labs 4* investigated whether involvement of the original researchers improved replicability—and unfortunately it did not.[11] And *Many Labs 5* replicated several studies from the *Reproducibility Project: Psychology* again, to assess possible reasons for failed replications that were put forward by the original authors; addressing these issues did not appear to improve replication rates...[12]

Before I clarify what exactly these worrisome results mean, let's consider how other disciplines fare when it comes to reproducibility.

9 See OPEN SCIENCE COLLABORATION, 2015.
10 See Ebersole et al., 2016.
11 See Klein et al., 2022.
12 See Ebersole et al., 2020.

Text Box 1.1

Ioannidis introduces the Positive Predictive Value (PPV), which is the conditional probability that if a study shows a statistically significant result, there is actually a true effect underlying it. In other words, it is a measure of how often significant positive results are actually true positives. This percentage is affected by four factors:

a. Pre-study prior (R), the probability that a tested relationship is true even before testing;
b. Statistical power ($1-\beta$), the probability of finding supporting evidence if the assumed relationship is true;
c. Type 1 error rate (α), the probability of finding supporting evidence if the assumed relationship is false;
d. The bias rate (u), the extent to which results are biased to be positive, either through publication bias or active manipulation.

In the absence of bias, PPV equates to:

$$\frac{(1-\beta)R}{(1-\beta)R+\alpha}$$

Bias, however, will increase the total number of positive observations, such that PPV reads:

$$\frac{(1-\beta)R+u\beta R}{(1-\beta)R+u\beta R+\alpha(1-u)+u}$$

The table below presents the PPVs for the different scenarios outlined in the main text.

Statistical power	Pre-study prior	Type 1 error rate	Bias	PPV
60%	50% (1:1)	0.05	0	0.92
60%	10% (1:9)	0.05	0	0.57
60%	10% (1:9)	0.05	30%	0.19

The replication crisis—not restricted to psychological science

Although in the last ten years the field of psychology has engaged with replication issues more than any other field of research,[13] the same problems have surfaced elsewhere. Let's have a look at biomedicine first. Alarmed by a report that the success rates of clinical phase II trials had fallen from 28% to 18% in recent years,[14] and by in-house rumours about the low reproducibility of clinical trials in general, researchers from Bayer Healthcare attempted to replicate the results of sixty-seven published studies (primarily in oncology).[15] They were successful in *less than 25%* of the cases. At around the same time, researchers from Amgen and the University of Texas reported an even lower percentage: their research departments had been able to replicate only 11% of fifty-three landmark studies in the field.[16] Later, in 2021, *The Reproducibility Project: Cancer Biology* published the findings of its re-examination of fifty-three high-impact cancer biology papers published between 2010 and 2012. Although the consortium achieved a higher replication success rate (46%), its results were shocking in another sense: replication effect sizes were *85% smaller on average* than the original findings.[17]

Economics came relatively late to the replication game, but by now that discipline too has been shown to seriously suffer from reproducibility issues. A study that tried to replicate the findings of experiments published in two leading economics journals (viz., the *American Economic Review* and *The Quarterly Journal of Economics*) managed to replicate only 61% of the original findings.[18] Another paper found conflicting results in 20% of the examined cases.[19] And yet another study, concerning findings published in the reputable *Strategic Management Journal*, established conflicts between original studies and replications in 34% of cases.[20] Lastly, starting from the same datasets

13 For an overview, see Nosek et al., 2022; Stevens, 2017.
14 See Arrowsmith, 2011.
15 See Prinz et al., 2011.
16 See Begley & Ellis, 2012.
17 See Errington et al., 2021.
18 See Camerer et al., 2016.
19 See Goldfarb, 1997.
20 See Bergh et al., 2017.

used in the original studies, researchers in macro-economics managed to reproduce only 32% of the original results, and the success rate improved only modestly when the help of the original researchers was solicited.[21] (Want more proof that economics is in rough shape? Check out the references in this last footnote—there's plenty more there.)

Also in the social sciences, a replication study of twenty-one experiments published in *Science* and *Nature* (2010–2015) could not detect the same effect, let alone the same effect size, as in the original study in 38% of cases.[22] In yet another study, 226 social science researchers evaluated 3,000 scientific claims and provided a quantitative assessment of how likely each claim was to hold up in an independent replication.[23] Likelihoods were soberingly low, and ranged from 58% in economics to as low as 42% in psychology. Disgusted by what he had seen, one of the assessors described the body of literature he had gone through as "a vast landfill—a great agglomeration of garbage extending as far as the eye can see, effluvious waves crashing and throwing up a foul foam of p=0.049 papers".[24] A grim view, indeed.

If you think the replication crisis is limited to social science and biomedicine, I need to disappoint you. A massive replication study in marine biology found no effects of ocean acidification on the behaviour of fish, despite a decade of research suggesting otherwise—suggesting, more particularly, that acidification disorients fish such that they swim towards their predators.[25] A study on 1,989 articles in water resource management estimated that only up to 6.8% of the findings might be replicable, even on the condition that the original articles provided enough information to allow replication.[26] Similar issues were identified in geoscience and in research on machine-learning.[27]

21 See Chang & Li, 2015. For economics, see also Bruns et al., 2024; Ferraro & Shukla, 2020; Huntington-Klein et al., 2021. For management, see the overview in Hensel, 2021; Pagell, 2021. For finance, see the overview in Pérignon et al., 2024. For marketing, see Trafimow et al., 2020. For innovation science, see Sarstedt et al., 2024.

22 See Camerer et al., 2018.

23 See Gordon et al., 2020.

24 See https://fantasticanachronism.com/2020/09/11/whats-wrong-with-social-science-and-how-to-fix-it/. For political science, see the overview in Brodeur et al., 2024. For criminology, see Lösel, 2018. For education, see Frias-Navarro et al., 2020.

25 See Clark et al., 2020.

26 See Stagge et al., 2019. For other papers in biology, see Gould et al., 2025; Kimmel et al., 2023. For neuroscience, see Bruns et al., 2024; Miłkowski et al., 2018.

27 See Ferrari Dacrema et al., 2019; Kapoor & Narayanan, 2023; Konkol et al., 2019.

Important fields of physics, too, are grappling with a replication crisis.[28] In recent years, several widely publicized breakthroughs—particularly in quantum computing and superconductivity—were published in top-tier journals like *Science* and *Nature*, only to be retracted after other researchers failed to reproduce the results. The retractions have caused such deep concern in the physics community that a broad coalition of physicists, journal editors, and NSF representatives convened an emergency meeting in Pittsburgh to hatch out the best way forward—to deal with what is, according to the organizers of the conference, just the tip of the iceberg.[29]

In engineering, experiments that go into patent filings appear extremely hard to replicate.[30] Sample sizes tend not to be disclosed, experiments are often not blinded, protocols suffer from a lack of randomization and even from a lack of statistical analysis. This means that a large portion of patents likely do *not* work.

Engineering also scores poorly in a survey study conducted, quite ironically, by *Nature* (among 1,576 scientists in total): as many as 69% of the engineers who participated in the survey indicated that they had tried and failed to reproduce another engineer's experimental results.[31] The numbers for researchers from other disciplines look just as poor: 87% of chemists had tried but failed, 77% of biologists, 69% of physicists, 67% of medical researchers, 64% of earth and environmental scientists, and 62% of all others. Additionally, more than 50% of the respondents were said to have failed to replicate their own findings.

The same survey study is alarming in yet another sense: most replications, even successful ones, tend not to get published. To put this into perspective, consider Figure 1.1, which presents the answers of respondents to the question "Have you ever tried to publish a reproduction attempt?" When it comes to successful replications, only 24% of respondents indicated that they had tried and in actuality had managed to publish their replication; 12% had tried but failed. This means that *76% of respondents' successful replications have not been reported on in any kind of publication* (i.e., the 64% of respondents who didn't even

28 See Chen, 2024; Padavic-Callaghan, 2024.
29 See Padavic-Callaghan, 2024.
30 See Freilich, 2020.
31 See Baker, 2016.

try to publish, plus the 12% who made an unsuccessful publication attempt). For unsuccessful replications, this number is even higher, viz., *87%*. In sum, although replication attempts appear common, they are severely under-reported; what we see in journals is just the tip of a gigantic epistemic iceberg.

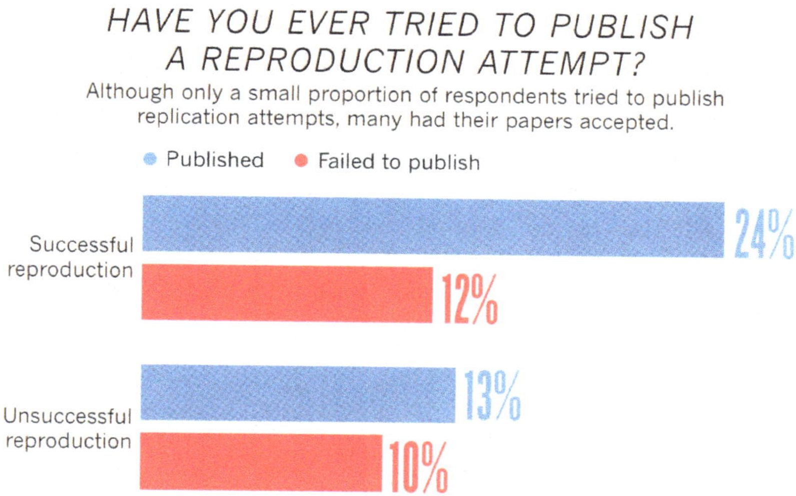

Fig. 1.1. Survey answers to the question "Have you ever tried to publish a reproduction attempt?" Baker (2016).

What does it all mean?

These results sound alarming indeed. For scientific knowledge to be reliable and robust, at *minimum*, experiments repeated under the same conditions should produce the same results.[32] Indeed, as the philosopher of science Karl Popper famously put it, "single occurrences that cannot be reproduced are of no significance to science".[33] But, apparently, many of science's findings are of the single-occurrence type. Science, thus, doesn't appear as reliable as we thought it was. Let's get clear on what exactly this means.

First and foremost, it means that *we don't know what we thought we*

32 See Romero, 2019.
33 See Popper, 1959, p. 66.

knew. Consider the 100 experiments that the *Reproducibility Project: Psychology* attempted to replicate. Before the replication attempt, we might, perhaps somewhat naïvely, have thought that the results of these experiments formed a body of robust knowledge—there was little *a priori* reason to doubt the studies' reliability. The *Reproducibility Project: Psychology*, however, gave us such a reason, and a strong one at that: the project's replication rate of only 36% took away much of our confidence.

But we need to be careful here. It would be far too hasty to conclude that unsuccessful replications demonstrate flaws in the original studies. For diverging results could be attributable to a variety of factors. Flaws in the original are one such factor. Flaws in the replication are another. Then there is random variation and subtle, perhaps non-codifiable, differences in the way the originals and replications were carried out. Or, as Lakens puts it:

> When a finding fails to replicate, logically there are three possible reasons for a non-significant result. First, the original result could be a false positive, or Type 1 error. Second, the replication study yielded a false negative, or a Type 2 error. And third, the replication study might yield a non-significant result because there is an unknown moderator: Under some conditions the effects can be replicated reliably, while under other conditions the effect fails to emerge, but researchers do not yet understand what those conditions are.[34]

So rather than concluding the falseness of the original and the truth of the replication, it is much safer to say that, in the case of failed replications, *we simply don't know what to believe about the hypotheses in question*.

Note that it is also far too simple to conclude the truth of the findings that *were* successfully replicated. Successful replications *do* increase our confidence in our findings. Still, any alternative explanation for the original finding also qualifies as an alternative explanation for the replication.

Thus far, we have been concerned only with the status of our beliefs regarding the studies that were re-assessed in the *Reproducibility Project: Psychology*. But the results of the project also affect our beliefs regarding other studies in psychology. For the 100 studies in the *Reproducibility Project* were selected more or less randomly; accordingly, the project plausibly unearthed a general pattern, implicating more than just

34 See Lakens, 2023, pp. 10–11.

the 100 studies in the sample. By way of illustration, suppose the *Reproducibility Project* had yielded 100% successful replications. And suppose we were to assess our confidence in the findings of a novel study not in the sample. The random sampling would, defeasibly, justify acceptance of the findings, even without replication effort. Given the actual replication rate of the *Reproducibility Project*, however, we don't know whether to accept or reject. For any given finding, we would need several replications to settle the issue. But the vast majority of findings in psychology have not been subject to replication, let alone intensive replication—or perhaps they have, but this has gone unreported. Thus, it is hard to tell what to make of the vast majority of findings in psychology. They might be right, they might be wrong; we can't tell. In other words, except in the case of studies that exhibit, say, serious methodological flaws, we don't really *know what we know* (*and what not*).

This point extends to (sub-)disciplines that have not undergone systematic replication efforts. Before the replication crisis, the burden of proof lay with those alleging a reproducibility problem in a given field. In light of what we now know—namely, that every discipline subjected to large-scale replication tests has revealed substantial issues—that presumption has shifted. Without compelling evidence to the contrary, we have reason to suspect that any empirical discipline may face similar challenges. The default stance, therefore, can no longer be one of trust but of epistemic caution. In short, even in disciplines that have not been tested, we do not know what we know (or what we do not know).

The replication crisis also has broader implications. Obviously, theories and practical applications that are informed by experimental work now appear to stand on fairly shaky ground—in psychology and elsewhere. Also, the crisis may affect the public perception of and trust in science.[35] And this is worrisome, indeed, because, as Anvari and Lakens put it:

> There is a social contract between science and society, where in return for the public's support, science is required to transparently produce reliable knowledge about how the world operates.[36] No single person [among the general public or otherwise] has the time to become an expert in all scientific fields, and therefore everyone must have some level of trust in scientific findings, scientists, or the scientific community as a whole.[37] For

35 See Anvari & Lakens, 2018.
36 See Gibbons, 1999.
37 See Hendriks et al., 2015; Pittinsky, 2015.

> these reasons, if we want science to continue to play an important role in knowledge generation for society, to continue to be supported by and to receive public funding, we must be concerned with how science is perceived by the general public.[38]

For all the above reasons, it is fair to call the replication crisis a crisis.[39] And it is not the only one.

The replication crisis—not the only crisis

For scientific knowledge to mature, researchers need to engage in several distinct epistemic activities (see Figure 1.2). Research prototypically starts—and, unfortunately, often ends—with the *observation and identification of new phenomena*, e.g., the tendency of subjects to walk out of the lab more slowly after having been exposed to words relating to old age.

Ideally, researchers subsequently assess the *reliability and internal validity* of these findings. Reliability refers to the consistency of a measure. It tells us whether the results can be reproduced under the same conditions, and is assessed by means of direct replication. Doyen et al.'s step-by-step iteration of Bargh et al.'s old-age priming study is an example of this (as are all the studies in the *Reproducibility Project: Psychology*). Just because a finding can be reproduced, however, that doesn't mean it is correct. A mis-calibrated thermometer may, under the same conditions, consistently give the same result (i.e., be reliable), but, due to mis-calibration, also be consistently wrong (i.e., have low validity). More generally, internal validity has to do with accuracy: it tells us how well an instrument measures what it is supposed to. The way a study was designed, conducted, and analysed impacts internal validity; improper randomization or inadequate statistical testing, for instance, undermines a study's internal validity. The methodological discussions prompted by Bem's pre-cognition studies were in effect

---

38 See Anvari & Lakens, 2018, p. 266.

39 Some authors say that the word 'crisis' is overblown. Stroebe & Strack (2014), for instance, talk about the "alleged 'replication crisis'". They point out that exact replications (such as the ones attempted in, e.g., the *Reproducibility Project: Psychology* and the *Reproducibility Project: Cancer*) are "not sufficient" (p. 59), for they only test the reliability of experimental procedures. I agree. We also need non-exact (or so-called 'conceptual') replications, that is, replications testing ideas by means of new experiments using different experimental variables. But that such additional work is needed doesn't make the failure to exactly replicate a finding unproblematic—we want our science to use reliable experimental procedures, no?

discussions about the internal validity of widely used psychological protocols and statistical methods.

The original findings can be put on an even more robust empirical footing by carrying out *external validity assessments*. Such assessments determine the extent to which findings can be generalized to other contexts. Bem's and Bargh et al.'s original experiments were carried out on subjects from WEIRD (Western, Educated, Industrialized, Rich, and Democratic) societies. Given that WEIRD people are among the least representative populations one could find for generalizing about humans,[40] a reasonable external validity check would involve re-running the experiments on non-WEIRD subjects. Another reasonable external validity check concerns generalization to real-life settings. The idea here is to study, say, priming 'in the wild', in order to ascertain that priming is a genuine phenomenon rather than an artifact of laboratory conditions. One of the principal reasons for carrying out such assessments is that it increases the applicability or relevance of findings—e.g., if indeed a genuine phenomenon, priming could be exploited to nudge people to behave in desired ways.

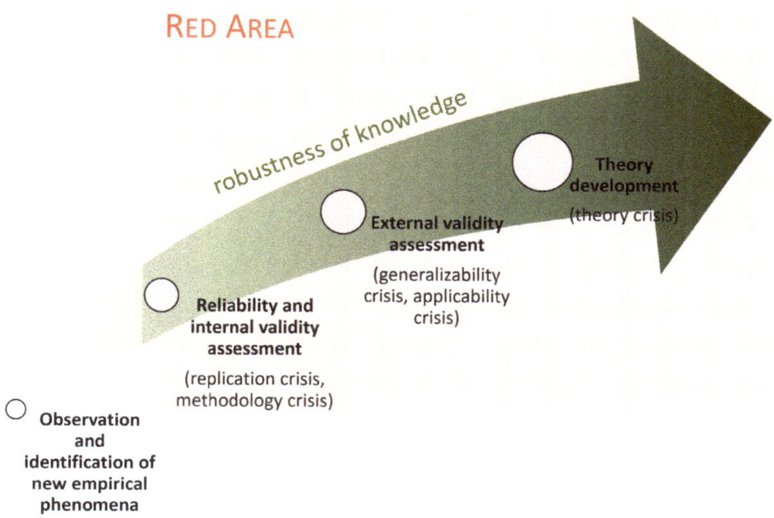

Fig. 1.2. A four-pronged model of scientific activity (with associated crises), with the arrow indicating increases in the robustness of knowledge. Image by the author.

40 See Henrich et al., 2010.

Finally, knowledge may mature through *theory development*. Theories, for one, may explain and unify otherwise isolated observed phenomena. So, if found to be robust, the results of Bargh et al.'s old-age priming study could perhaps be explained by a general theory regarding the automaticity of behaviour—a theory that would pertain to all sorts of priming, not just the type of priming studied by Bargh et al. Further, theory development may aid the prediction of new empirical phenomena—such that we start again in the lower-left corner of Figure 1.2, resulting in a cyclical figure. And, relatedly, it could help in the clarification of concepts necessary for studying those new (but also old) phenomena.

The reason why I introduce this rough four-stage model of scientific activity—which, admittedly, may gloss over some of the details—is that it helps me to make the following point: with each activity that could increase the robustness of the findings of the first activity (viz., *observation and identification of new empirical phenomena*), there is one or more associated crisis in important parts of science. Insufficient emphasis on *reliability and internal validity assessment* has engendered not just a replication crisis but also a methodology and measurement crisis; insufficient emphasis on *external validity assessment* has led to a generalizability and an applicability crisis; and insufficient emphasis on *theory development* has prompted a theory crisis.

As for the *methodology and measurement crisis*, the replication crisis has engendered a lively debate about problematic methodological practices and about ways of improving practices—within and outside of psychology.[41] Also, the replication crisis has already led to positive changes: large-scale consortia have enabled the collection of much larger samples of data, preregistration of research is becoming more common, scientists have adopted more sophisticated statistical techniques, and so forth. But even though the sense of urgency might have dissolved, it is too early to say that the methodology and measurement crisis is over. Many disputes haven't been settled, and the prospects of them soon coming to a close look dim: e.g., there is still no consensus on whether to use a frequentist or Bayesian framework for hypothesis testing, nor on the proper threshold for statistical significance (0.05 or 0.005? Or

41 For psychology, see LeBel & Peters, 2011; Wagenmakers et al., 2011. For other disciplines, see, e.g., Button et al., 2013.

should we give up on statistical significance testing?), nor on whether researchers should preregister their research, nor if peer reviews should be open and signed or private and anonymous, nor on whether change needs to emerge bottom-up or be introduced top-down.

In this context, I fear I have to open yet another can of worms, and discuss a class of studies that are often referred to as 'many-analyst studies'. These are investigations in which the same dataset is analysed independently by a large number of research teams. Would independent analysts, working on the same data to address the same hypothesis, reach similar conclusions? One would hope so. But, it appears, often they do not. Consider the sociological study entitled "Observing many researchers using the same data and hypothesis reveals a hidden universe of uncertainty", published in the reputable journal *Proceedings of the National Academy of Sciences*.[42] Seventy-three research teams were involved. They used the same data to independently test the same prominent social science hypothesis: that greater immigration reduces support for social policies among the public. *No two teams appeared to arrive at the same numerical results*—teams varied widely regarding the conclusions they drew, despite identical starting conditions. Neither researchers' expertise nor their prior beliefs or expectations explained the striking variability in outcomes. In fact, more than 95% of the variance in numerical results remained unaccounted for. This suggests that much of the divergence stems from tacit, non-codifiable differences in analytical choices across teams.

Note that the authors of the article explicitly do *not* talk about research teams performing the *correct* (or even best) analysis: given the wide variance of results, and given the expertise and high motivation of the teams, there is little to go on to determine what's right or best. What's more, because so many research decisions are tacit and non-codifiable, we cannot confidently design interventions that will reliably make things better.

And, again, evidence of profound methodological uncertainty is not confined to sociology, economics,[43] and psychology.[44] By now, many-

42 See Breznau et al., 2022.
43 See Huntington-Klein et al., 2021; Sarstedt et al., 2024; Schweinsberg et al., 2021..
44 See, e.g., Coretta et al., 2023; Silberzahn et al., 2018.

analyst studies have also been conducted in fields such as neuroscience,[45] ecology,[46] and epidemiology[47]—yielding broadly similar results. Taken together, these findings shift the burden of proof onto fields that have not yet undergone such scrutiny: it is now up to them to demonstrate that their methodological house is in order.

The article that probably has contributed most to scholars in the social and behavioural sciences talking about a *generalizability crisis* is a paper I already referred to, that is, Henrich et al.'s 2012 article "The WEIRDest people in the world?".[48] In that paper, the authors showed that WEIRD people form the bulk—up to 96%—of research participants in behavioural science (including experimental psychology, cognitive science, and economics) and that, based on observations of these WEIRD subjects, researchers tend to generalize findings to all humans. They went on to show that, given the admittedly sparse evidence about non-WEIRD subjects—who form the bulk of the human species—such generalizations cannot be justified. WEIRD people are really weird: *they*, rather than non-WEIRD subjects, are the outliers when it comes to visual perception, fairness, cooperation, spatial reasoning, categorization and inferential induction, moral reasoning, reasoning styles, self-concepts and related motivations, and the heritability of IQ; accordingly, it is reasonable to assume that *they*, rather than non-WEIRD subjects, will be the outliers in other domains. Henrich et al.'s paper, then, was a plea to incorporate, as much as possible, non-WEIRD subjects in the study of human nature.

Despite efforts—primarily of a small number of newly established large consortia (see also Chapter 5)—to address the worries of Henrich et al., the crisis doesn't seem to be over. Some, such as Tal Yarkoni, express frustration that the replication crisis has resulted in too many resources being spent on reproducibility at the expense of generalizability.

> At the same time, the current focus on reproducibility and replicability risks distracting us from more important, and logically antecedent, concerns about generalizability. The root problem is that when the manifestation of a phenomenon is highly variable across potential

45 See Botvinik-Nezer et al., 2020.
46 See Gould et al., 2025.
47 See Kowall et al., 2025.
48 See Henrich et al., 2010.

> measurement contexts, it simply does not matter very much whether any single realization is replicable or not.[49]

Also outside the social and behavioural sciences, generalizability remains a major concern. It is well known that women and the elderly have been under-represented in biomedical randomized control trials (RCTs); they remain so even after regulatory changes by such bodies as the U.S. Food and Drug Administration and the European Union.[50] But the problem has proven to be broader: many other salient patient characteristics (demographic- and disorder-specific, factors relating to functioning and comorbidities, and behavioural, environmental, and equity factors) continue to be ignored in RCTs reported in the four leading general medical journals (*BMJ*, *JAMA*, *The Lancet*, and *The New England Journal of Medicine*).[51]

The recent creation of large research consortia specifically aimed at tackling generalizability concerns (again, see Chapter 5) shows that these problems also affect fields such as primatology, biology, and ecology. In fact, any field facing a replication crisis is almost certainly facing a generalizability crisis as well. After all, if results cannot be reproduced under the very same conditions, it is even less likely they will hold under different ones. Think of a patent: if an invention fails under the conditions specified in its patent file, it would be nothing short of a miracle for it to work elsewhere. The same logic applies to unreproducible findings in physics, chemistry, and so forth.

The *applicability crisis* refers to the fact that scientific findings often remain too far removed from real life (e.g., they pertain to highly idealized lab settings, they answer 'irrelevant' questions) to be practically applicable and to inform policy. It is difficult to establish the existence of this crisis: success stories may be easy to spot, but harder to judge when it comes to their *relative* prevalence.[52] In psychology, anecdotal evidence comes from judgments of well-respected figures,[53] and from conversations with fellow psychologists;[54] these sources repeatedly

49 See Yarkoni, 2022, p. 12.
50 For a review of the evidence, see Daitch et al., 2022.
51 See Malmivaara, 2019.
52 See Giner-Sorolla, 2019.
53 Ibid.
54 See Lakens, 2023.

raise concerns about applicability. Berkman and Wilson give some more objective evidence.[55] Surveying the publication criteria of the top ten psychology journals, they found that very few journals aimed their articles to reach out to audiences beyond psychology. Further, Berman and Wilson found that very few articles published in the most cited journal in social psychology—i.e., the *Journal of Personality and Social Psychology*—met generally accepted applicability criteria. The authors conclude:

> Our casual observation is that psychological theory has become unmoored from the guiding principle of practicality and is drifting toward more nuanced or myopic theoretical questions that are less relevant to helping solve the problems that people care about.[56]

In biomedical research, too, there is a tendency to focus on 'irrelevant' or 'wrong' questions. For one, if biomedical research addressed the most pressing health issues, there should be a correlation between a society's public funding of research and its disease burden; many studies find only a weak correlation, if at all.[57] For another, there are dramatic mismatches between the priorities of researchers and those of patients and clinicians.[58] For example, a study by Tallon et al. found that while only 9% of patients with osteoarthritis prioritized research on drugs—most patients prioritized research on physiotherapy, surgery, and educational and coping strategies—over 80% of RCTs were drug assessments.[59] Also, a large fraction of clinical studies turn out to be of limited use due to poor design: studies are often too small or too short to yield robust insights, fail to reflect real-world conditions, and neglect to situate themselves properly within existing evidence.[60]

Regarding applied science and engineering, the number of patents issued each year has been soaring since the late 1980s (for why this is so, see Chapter 3). But, at the same time, the quality of patents is going down, in the sense that the proportion of uncited patents has

55 See Berkman & Wilson, 2021.
56 Ibid., p. 869.
57 See Chalmers & Glasziou, 2009, and references therein.
58 Ibid.
59 See Tallon et al., 2000.
60 See Ioannidis, 2016.

dramatically increased.[61] Apparently, applied scientists and engineers produce an increasing amount that never gets used—either because it doesn't work as intended or because it doesn't meet any real demand.

The final crisis concerns *theory development*. The worry here is that science accumulates facts, without these facts being integrated into theories—theories that not only allow us to unify otherwise isolated phenomena, but also to predict new phenomena from general premises, to causally explain phenomena, and so forth. Researchers, so the complaint goes, have been obsessed with methodological virtuosity (e.g., sophisticated statistical testing) and 'fun' studies at the expense of the formulation of precise theories.[62] And, insofar as they *do* develop theories, researchers are reluctant to use any theory other than their own. It is as if theories are like toothbrushes—no self-respecting person wants to use anyone else's.[63] This leads to a proliferation of theories, none of which are decisively accepted or rejected. So, for example, the book *ABC of Behavior Change Theories*[64] lists no less than eighty-three theories in the field of behaviour change alone[65]—each theory 'hanging around' without conclusive evidence for or against it.[66]

Theory development, or the lack thereof, is openly discussed in psychology and ecology,[67] much less so in other disciplines, if at all. This might mean that other disciplines do better than psychology and ecology, or that the problem has not been recognized yet. Surely, methodological sorcery and studying 'sexy' topics do not seem endemic just to psychology and ecology. But the mere presence of such practices in a discipline does not imply that theory building has been abandoned by that discipline.. The best evidence for a theory crisis outside psychology is circumstantial and comes from a bibliometric study by Zeng et al.[68] These authors looked at the publication data of all journals of the American Physical Society (1893–2010; 482,566 articles) and determined the rate at which scientists switched between research topics. That rate,

61 See Gandal et al., 2021.
62 See Eronen & Bringmann, 2021; Lakens, 2023, and references therein.
63 See Mischel, 2008.
64 See Michie et al., 2014.
65 See https://x.com/lakens/status/1174963097158578176
66 See Eronen & Bringmann, 2021.
67 See, again, Eronen & Bringmann, 2021; Lakens, 2023. For ecology, see Nakagawa et al., 2025.
68 See Zeng et al., 2019.

they found, has considerably increased over time. Scientists, thus, seem to increasingly spend time on exploration (discovery of new ideas) and decreasingly on exploitation (the further development of ideas). Or, in terms of Figure 1.2: they devote less time to activities within the RED AREA—including theory development.

To be sure, the argument in the remainder of this book does not hinge on the existence of a theory crisis in every corner of science. Our case does not require universal collapse. What matters is that crises—whether of replication, methodology, generalizability, applicability, or theory—have emerged in an unsettling number of domains. If cracks appear in so many places, it becomes difficult to simply assume that other areas are immune. The point is not that science as a whole is broken, but that the old default—widespread reliability—no longer looks safe.

To make matters worse: Each crisis in fact harbours two crises...

In the above section, I have glossed over a distinction that will become important for the remainder of our story: each crisis in fact harbours two types of crises. Let's refer to them as A-crises and B-crises—such that there is an A-replication crisis and a B-replication crisis, an A-generalizability crisis and a B-generalizability crisis, and so forth.

An A-crisis refers to a crisis in confidence in published results based on empirical evidence from studies that have actually been conducted. So, for instance, the crisis that the ninety-seven studies of the *Reproducibility Project: Psychology* gave rise to is an A-crisis: the results of the studies that were actually carried out (i.e., the ninety-seven replications) decreased our confidence in the robustness of the findings of the original studies. In short, A-crises are triggered by what we *do* know—by data that actively undermine prior claims.

A B-crisis refers to a crisis in confidence in published results in light of the fact that relevant studies have *not* been carried out. The B-replication crisis, thus, simply refers to the fact that, generally, researchers in psychology spend too little time on replication—such that, generally, our confidence in findings in psychology should be low. Unlike A-crises, B-crises stem from what we *don't* know—from the absence of robustness-checking studies that would allow us to assess the reliability of published claims. B-crises thus are crises that result from the fact that scientists insufficiently do what they should do, namely engaging in robustness-increasing activities (the activities in the RED AREA in Figure 1.2). They reflect a structural neglect of epistemic maintenance work.

Conclusion

Science will never be perfect. But the evidence reviewed in this chapter suggests that the problem is not just a handful of flaws—it runs deeper. Across multiple domains we see recurring crises: crises of replication, methodology, generalizability, applicability, and theory. Psychology, economics, cancer biology, ecology, neuroscience, and even parts of physics, chemistry, and engineering have all faced serious challenges. To be sure, the evidence for such crises is not equally strong everywhere; some disciplines may appear relatively stable. Yet given how widespread these problems have proven to be, the onus now lies with fields that have not been scrutinized to demonstrate that their methods and findings deserve our trust.

What, then, explains this poly-crisis? Why do so many scientific fields—despite their differences—grapple with similar foundational issues? The next chapter argues that the root cause lies in a systematic neglect of a particular kind of scientific work: the kind that falls into the RED AREA of Figure 1.2. It explores how prevailing incentives have steered researchers away from this zone, and how this collective drift has led us into the current predicament.

2. Neomania as an explanation of the poly-crisis

Apparently, science is going through a poly-crisis—but, how come?

According to the standard story, the primary culprit is academia's 'publish or perish' culture. Scientists fiercely compete for (ever more) limited resources—jobs, promotions, bonuses, funding, awards, accolades. The more proficient in research, the more likely a scientist is to stand out in that competition. Now, a common currency for a scientist's research proficiency is their publications. So, unless a scientist publishes a lot, they will proverbially perish in the academic rat race. And to publish a lot is easier if one rushes their research and cuts corners—*voilà*, the low quality of published findings.

The idea that academia's 'publish or perish' culture might be responsible for an inverse relationship between research quality and research quantity is hardly new. It has been with us at least since 1928, when the sociologist Clarence Marsh Case wrote:

> If it be true that, for the time being at least, the quality of American sociological writing is in inverse ratio to its quantity, the reason is to be sought, among other things, in the fact [...] that the system of promotion used in our universities amounts to the warning, "Publish or perish!".[1]

To this old explanation has recently been added another, one in terms of publication bias. In this account, scientific journals are biased towards publishing *positive* results (i.e., results that a statistically significant effect exists); negative results (i.e., results that no statistically significant effect exists) have a much harder time surviving the

1 See Case, 1928, p. 325.

 https://doi.org/10.11647/OBP.0507.02

rejection mills of, say, *Nature* and *Science*.[2] On the face of it, journals' bias towards positive findings doesn't by itself seem to explain, say, low reproducibility rates. After all, in principle any positive finding might be backed by the best possible evidence, including evidence from unpublished but successful replications. Yet, as we have seen in the previous chapter, it can be shown that, even if researchers work diligently, a large proportion of published positive findings may be false. Obviously, things get worse if researchers, in an attempt to obtain positive and therefore publishable results, start cutting corners. For instance, researchers might conduct studies with low statistical power (see also Chapter 1). Or they might engage in the questionable research practice of data-dredging, which refers to exhaustively testing hypotheses with the same dataset and reporting only significant results. Note the reason why such exhaustive testing substantially increases the odds of getting positive results. With a significance level of $\alpha = 0.05$, there is a 5% chance of a false positive when in reality there is no effect. So if a researcher tests 100 hypotheses that have no real effect, we should expect about 5 false positives just by chance. If the researcher then reports only those 5 'significant' results and ignores the rest, it will look as if real discoveries were made—even though all of them are false. This is why data-dredging can generate impressive-looking findings out of pure noise. Other practices that may lead to low reproducibility include stopping data collection when a positive result is found, excluding outliers, poor or selective reporting, and outright fraud (e.g., data fabrication).

Enter neomania

Together, 'publish or perish' and publication bias seem to go a long way in explaining the low reproducibility, generalizability, and applicability of published findings, the poor methodologies being used to obtain such findings, and the poor quality of the theories that are derived from them. In other words, 'publish and perish' and

2 Less commonly, publication bias also includes ignoring undesirable significant results (e.g., iatrogenic effects) and being more lenient in publishing non-significant results.

publication bias account reasonably well for 'A-crises', as I called them in Chapter 1. However, they leave B-crises unexplained. Indeed, 'publish or perish' and publication bias do not *per se* militate against replicability/generalizability/applicability studies, methodological studies, or theory development studies; for, even in a regime of 'publish or perish' and of strong publication biases, a positive replication/generalization/application/methodological/theoretical study is as publishable as a study reporting on a positive *novel* result. Yet, this is not what we see: we see an over-representation of studies of the latter type or, conversely, an under-representation of studies of the former type. In graphical terms, too little work is done in the RED AREA in Figure 1.2 in Chapter 1.

In order to explain these B-crises, we need to add neomania—an obsession with the new—to the mix. Journals, particularly high-ranking ones, are most keen on publishing not just positive, but positive *novel* results—that is, results pertaining to new empirical phenomena such as the therapeutic effect of a newly discovered substance, a hitherto never documented property of a sub-atomic particle, a new stimulus-behavioural response pair, a new macro-economic trend, a novel way of producing a (novel) technical artefact, and so forth.[3] Indeed, in a large scale study investigating the association between novelty and the acceptance for publication in forty-seven prominent science journals, Teplitskiy et al. (2022) found that higher novelty is consistently correlated with higher acceptance.[4]

Authors seem to pick up on this. Analysing all abstracts that appeared on PubMed between 1974 and 2014, researchers found an up to nine-fold increase in the use of words expressing novelty ("novel", "innovative", "unprecedented", and so forth; see Figure 2.1).[5] There is little reason to think that this pattern reflects a real improvement—i.e., that researchers in 2014 were in actuality better at producing novel results than their colleagues in 1974. It is much more likely that authors have come to realize that, in order to get published, it helps to (over-)emphasize the uniqueness of their study.

3 See Nosek et al., 2012.
4 See Teplitskiy et al., 2022.
5 See Vinkers et al., 2015.

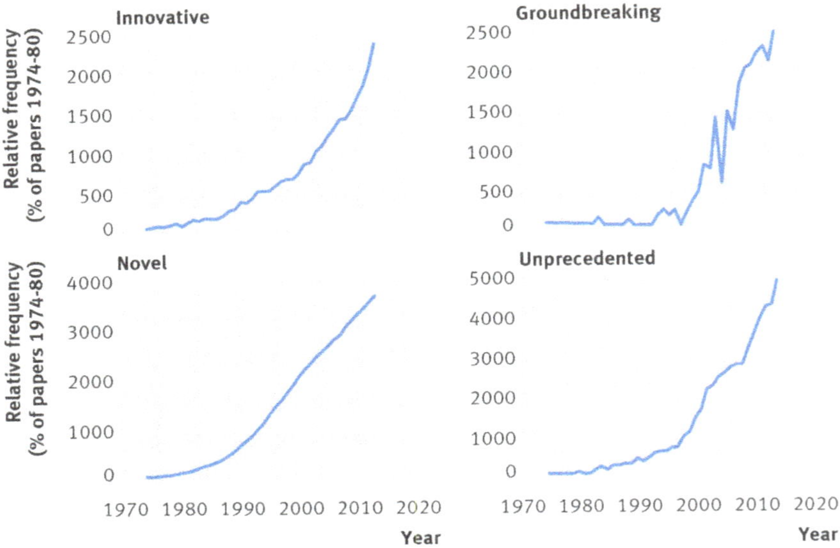

Fig. 2.1. Relative frequencies of the words "innovative", "novel", "groundbreaking", and "unprecedented" over time, as found in PubMed abstracts between 1975 and 2014. Adapted from Vinkers et al. (2015).

Given that research devoted to reproducibility/generalizability/applicability/theory development scores low on novelty (in the sense just described), such research generally has a hard time making it to the pages of (high-ranking) journals.[6] For researchers, it is thus rewarding not to replicate, not to test for generalizability, etc.—hence the B-crises. Indeed, as we have observed in the previous chapter, in a survey carried out by *Nature*, only a minority of respondents said they had ever tried to publish a replication study.[7] And several of those who did manage to publish a failed replication were prompted by editors and reviewers to downplay comparisons with the original study.

Note that neomania on the part of journals also reinforces my earlier explanation of the A-crises. The demand for novelty rewards researchers for being the *first* to report a finding. But being first often means rushing the research process—cutting corners in design, analysis, or validation—which in turn fuels A-crises. In sum, neomania contributes

6 To the extent that researchers engage in theory development, they might be accused of neomania too. See the toothbrush problem mentioned in Chapter 1.

7 See Baker, 2016.

not only to B-crises, but—together with 'publish or perish' pressures and the distorting effects of publication bias—it plays a central role in driving A-crises.

Interestingly, I am not the first to observe a tension between originality and rigor. Already in 1957, the sociologist of science Robert K. Merton observed such a tension.[8] Merton infers the priority rule from the numerous fierce disputes over priority in the history of science—e.g., Newton and Leibniz were in a long and painful controversy over the invention of calculus; Cavendish fought with Watt and Lavoisier about who had first demonstrated the compound nature of water; Laplace, Bernouilli, and Gauss were all at some point embroiled in quarrels about priority. Recognition is the principal reward for scientists, and recognition goes *exclusively* to those scientists who are the *first to discover a particular fact*. Accordingly, scientists are very strongly incentivized to do original work. Or, put much more eloquently,

> On every side the scientist is reminded that it is his role to advance knowledge and his happiest fulfillment of that, role, to advance knowledge greatly. This is only to say, of course, that in the institution of science originality is at a premium. For it is through originality, in greater or smaller increments, that knowledge advances. When the institution of science works efficiently—and like other social institutions, it does not always do so—recognition and esteem accrue to those who have best fulfilled their roles, to those who have made genuinely original contributions to the common stock of knowledge.[9]

Too strong an emphasis on original discovery, Merton continues, may lead to what we nowadays would call questionable research practices:

> Contentiousness, self-assertive claims, secretiveness lest one be forestalled, reporting only the data that support an hypothesis, false charges of plagiarism, even the occasional theft of ideas and, in rare cases, the fabrication of data,—all these have appeared in the history of science and can be thought of as deviant behavior in response to a discrepancy between the enormous emphasis in the culture of science upon original discovery and the actual difficulty many scientists experience in making an original discovery.[10]

8 See Merton, 1957.
9 Ibid. p. 639.
10 Ibid., p. 659.

Merton says that such questionable practices "have appeared in the history of science". Still, it is unclear how often he thinks they have and, accordingly, whether he finds the emphasis on originality healthy or not. Sometimes he writes as if the emphasis is kept in check by another institutional norm, namely the norm of humility. Humility "reduce[s] the misbehavior of scientists below the rate that would occur if importance were assigned only to originality and the establishing of priority".[11] It involves acknowledging the "heavy indebtedness to the legacy of knowledge bequeathed by predecessors" and "insisting upon [one's] personal limitations and the limitations of scientific knowledge altogether".[12] In other parts of the text, however, Merton suggests that "the culture of science [...] is pathogenic".[13]

If the emphasis on originality was unhealthy in 1957, what about now? Arguably, things have just gotten worse. For science has grown massively, with scientists yearly publishing about 2.5 million articles in thousands of journals. To stand out in such a flood of publications, humility is a lousy strategy—and Figure 2.1 attests to that. It is a lousy strategy also because, at the other end, journal policies incentivize rather than penalize intellectual arrogance. Bold novel claims are more likely to get cited. So journals may, and plausibly do, attempt to boost their impact factor (an unknown notion in Merton's time) by preferentially publishing such claims,[14] even though doing so may lead to more papers being retracted.[15] Additionally, high impact factor journals such as *Nature* and *Science* have an unforgiving word limit. They leave little room for humility and acknowledging uncertainty.

Generalized neomania

Universities

Neomania on the part of journals also *indirectly* leads universities and science funders to reward novelty, because traditional criteria—number of publications, citation counts, and journals' impact

11 Ibid., p. 646.
12 Ibid.
13 Ibid., p. 659.
14 See Hoekstra & Vazire, 2021.
15 See Fang & Casadevall, 2011.

factors—remain widely used in hiring and promotion decisions (at the level of universities) and in grant allocation (at the level of science funders).[16] Such criteria also inform decisions about pecuniary rewards. Financial bonuses for publishing in pro-novelty journals such as *Science* or *Nature* are sometimes substantive: individual researchers may receive from their employer up to US$5,000 in Australia, US$6,000 in the U.S., US$6,400 in the U.K., and a whopping US$165,000 in China.[17]

But universities, science funders, and in fact society at large also *directly* push researchers in the direction of novelty—science finds itself in a regime of *generalized* neomania. Let's start with universities.

First, universities expect their scientific staff to carve out a unique niche for themselves. It is not enough to be one among the many; researchers need to stand out in their disciplinary ecosystem. And what helps here is a willingness to explore—and if successful, to lay claim to—new areas of research.[18] By way of illustration, in order to become an associate or full professor, my own university asks candidates to conceive a unique label for the (unique) work they do. That is why I am currently called an associate professor in the Philosophy of Innovation—to the best of my knowledge, no one else in the world is. University students, too, are strongly encouraged to be novelty-seekers. For their bachelor's, master's, and, in particular, PhD theses, they typically need to carry out an original piece of research—independence of thought is considered to be more important an evaluation criterion than, say, proficiency in the assessment of existing bodies of knowledge.

Further, Merton's era of 'pure academic science' has been succeeded by an era of natural allyship between science and (technological) innovation, the era of what Etzkowitz dubs the "entrepreneurial university". The entrepreneurial university is governed like a firm, and works on the assumption that "science has emerged as an alternative engine of economic growth to the classic triumvirate of land, labor and capital, the traditional sources of wealth".[19] More specifically, it is, according to Etzkowitz,

---

16 See Rice et al., 2020.

17 See Abritis et al., 2017. I have no data for other countries, so am unable to assess how common the practice really is.

18 See https://www.science.org/content/article/how-establish-your-identity-scientist

19 See Etzkowitz, 2003, p. 109.

> an especially propitious site for *innovation* due to such basic features as its high rate of flow through of human capital in the form of students who are a source of potential *inventors*. The university is a natural incubator; providing a support structure for teachers and students to initiate *new* ventures: intellectual, commercial and conjoint. The university is also a potential seedbed for *new* interdisciplinary scientific fields and *new* industrial sectors, each cross-fertilizing the other.[20] (italics added)

Mind the italics in this quote: they perfectly illustrate contemporary universities' love for the new. Why, historically, they have developed this love is a story I will detail in Chapter 3. For now, what matters is that, in response to declining direct government investments in academic research, universities seek additional income; and that the monetization or commodification of *new ideas*—through patenting, creating spin-off companies, setting up university-industry partnerships and the like—has become a popular way of (attempting to) generate such income.[21]

Financial rewards are not the only incentive for novelty seeking. Universities are increasingly obsessed with university rankings, including the Times Higher Education ranking and Reuter's innovation ranking.[22] The obsession expresses itself in a global frenzy among governments and university administrators to make it to the top 100 of these lists—or at least not to perform worse than previous years. This is because the prestige associated with a high rank increases a university's global competitiveness when it comes to attracting excellent students, excellent scientific personnel, and excellent additional investments. Crucially, rankings are typically informed by universities' patent data (e.g., patent volume, patent success, citations to patents, etc.)—where patents need to be *novel* to get granted. In an attempt to end up high in the rankings, universities each year spawn thousands and thousands of patents—very often, as we saw in Chapter 1, of questionable quality.

And then there is the push to have one's research covered by the media. Dorothy Nelkin starts her *Selling Science: How the Press Covers Science and Technology* with an amusing but telling quote from Edwin E. Slosson, one of the first science journalists in the U.S., who, in 1924, remarked derogatively:

---

20 Ibid., p. 112.
21 See Zilwa, 2005.
22 See Hazelkorn, 2015, p. 1.

> The public that we are trying to reach is in the cultural stage when three-headed cows, Siamese twins and bearded ladies draw the crowds to the side shows.[23]

To provoke the interest of these crowds in science, Slosson suggests writing about science in short paragraphs ending in '-est':

> The fastest or the slowest, the hottest or the coldest, the biggest or the smallest, and in any case, the *newest* thing in the world.[24] (italics added)

In some respects, little has changed. Sensationalism remains a concern in science journalism.[25] And science news today also preferentially reports on the 'previously unknown'.[26] But other things *have* changed.

For one, since science budgets went down in the 1990s, scientists actively seek media attention, and rely on ever more sophisticated communication techniques to do so.[27] Key Performance Indicators (KPIs) pertaining to public outreach have gained importance in departmental and university-level research audits. My department, for instance, has two such KPIs: 'dissemination of results to the general public' and 'altmetrics', the latter being a measure of the total attention that a publication receives across social media, blogs, news outlets, etc.

Researchers are more likely to get promoted when their work is featured in prominent news outlets[28]—and in light of that, they might feel pressured to sensationalize. In fact, as a report of the National Academy of Sciences states, sensationalizing science is to the benefit of all stakeholders but not the public.

> [Scientific] journals are more likely to publish [researchers'] work if it is sensationalized because [scientific journals] are interested in increasing their revenue or status. Companies are more likely to be able to raise funds or have their stock price increase. Journalists are also more likely to be read or viewed when they sensationalize or overdramatize the discoveries of science.[29]

23 See Nelkin, 1995, p. 1.
24 Ibid.
25 See Dempster et al., 2022.
26 See Rosen et al., 2016.
27 See Nelkin, 1995.
28 See National Academy of Sciences, 2018.
29 Ibid.

Further contributing to this are trends within science communication in general: the speed of today's news production, fierce and global competition among traditional and non-traditional content providers, declining revenues from advertising and the corresponding pressure to generate 'click-bait', etc.[30]

This is not to say that all science communication in actuality is sensationalist—far from it. Rather, what I *do* want to point out is that all incentives perversely invite sensationalism—and that, accordingly, it is extremely hard to resist. So it *is* understandable that scientists are tempted to exaggerate their claims in the press releases they issue.[31] It *is* understandable that science journalists are tempted not to publicize the limitations, shortcomings, and funding information of the studies they cover.[32]

If current trends invite researchers to choose mediagenic topics, they *a fortiori* invite them to focus on novelty—for novelty is just one aspect of what it is to be mediagenic. So, again, neomanic researchers are put at an advantage relative to their more conservative peers.

Science funders

As a frequent reviewer of grant proposals, it's virtually impossible not to develop an allergy to words that express novelty. To show they are no laggards, grant applicants (including me) copiously emphasize that their planned approach is 'new', 'original', 'unique', 'innovative', 'unconventional', 'unprecedented', 'transformative', 'ground-breaking', 'revolutionary', 'disruptive', 'trailblazing', 'paradigm-shifting', 'game-changing', 'great-lacuna-filling', 'uncharted-territory-charting'—and, preferably, a combination of all the above.

It can be just as annoying at the other side of the table. One reason I have time to write this book is that my latest grant proposal was rejected. And it was rejected precisely because it was deemed insufficiently innovative. Mind you, the idea that I wanted to put to the test was two(!) years old. So in a 2021 paper I formulated (but did not test) a hypothesis and, two years later, in 2023, the proposal to test that very

30 See Dempster et al., 2022.
31 See Bratton et al., 2019.
32 See Mellor, 2015.

same hypothesis was already considered unoriginal—at least according to the referee report I received. Indeed, originality has been found to be among the most common assessment criteria in grant peer review.[33]

These things are in a sense understandable. They fit with how funding agencies brand themselves as beacons of innovation. "Innovation Anywhere, Opportunity Everywhere" is not the slogan of Apple or Meta, but of the National Science Foundation (the largest science funding agency in the U.S.).[34] The European Research Council says it acts to "promote *creativity* and *innovative* research".[35] According to the Flemish Research Council, "science is the driver of innovation";[36] it is thus no surprise that it focuses on funding "innovative research".[37] The Dutch Research Council (NWO), finally, "realizes quality and innovation in science".[38] It includes "groundbreaking" in its core values [*sic*], and aims to contribute to "pioneer[ing] and explor[ing] the boundaries of existing knowledge, applications and processes". On the positive side, in response to worries about low reproducibility rates, NWO launched a pilot scheme that funded replication studies only—the pilot ran from 2016 to 2019 and has recently been revived. But, conspicuously, NWO is the *only* research council I know of that has taken such an initiative; for all others, it seems replication is not worth the money.

Additionally, what I have said about universities also applies to science funders: as they face growing pressure to demonstrate responsible use of taxpayer money, funders too increasingly expect awardees to reach out to non-academic crowds—which is easier for novel than for, say, replicated results.

At this point, I need to deal with an obvious objection. Recent studies suggest that grant-giving bodies tend to be more risk-averse than they should be.[39] Am I not exaggerating the pro-novelty bias of science funders?

My response is twofold. Firstly, the bias that has been documented is not a bias against novelty as such, but a bias against *very high levels*

33 See Hug & Aeschbach, 2020.
34 See https://www.nsf.gov/
35 See https://erc.europa.eu/about-erc/erc-glance, italics added.
36 See https://www.fwo.be/media/wnbdlrr4/annual-2019.pdf
37 See https://www.fwo.be/en/about-fwo/who-is-fwo/
38 See https://www.nwo.nl/en/what-does-the-dutch-research-council-do
39 See Boudreau et al., 2016; Veugelers et al., 2022.

of novelty. So Veugelers et al.[40] found a bias against proposals from the top 1%(!) of applicants when ranked in terms of the novelty of their prior publications—meaning that we cannot infer much about the role of novelty in the evaluation of proposals from the bulk (99%) of applicants. Accordingly, Veugelers et al.'s results are perfectly consistent with a strong positive relationship between novelty and funding up to the 1% top cut-off point. By the same token, the bias against novelty reported in another study, by Boudreau and colleagues, is driven primarily by what they call "particularly highly novel" proposals.[41] *The authors even acknowledge the existence of the positive relationship I just hinted at*: they found that for proposals that were not *highly* novel, "evaluation scores were increasing with incrementally greater levels of novelty".[42]

To be clear, my point is not that novelty takes precedence over every other consideration in the present funding landscape. Funding decisions are influenced by a wide range of factors—some explicit, others implicit. For instance, research shows that innovations proposed by members of under-represented groups are more likely to be discounted compared to those from majority groups, and that these individuals face lower chances of securing academic positions.[43] Research also indicates that certain structural characteristics of contemporary funding programmes—such as short project timelines and uniform resource allocations—create greater obstacles for radical or high-risk innovation than earlier funding models did.[44] But these findings do not contradict the claim that funders exhibit a strong novelty bias; rather, they simply suggest that the bias operates 'imperfectly', allowing significant innovative potential to go untapped.

Secondly, the above studies were carried out against a distinctly neomanic background—which exactly proves my point about the ubiquity of neomania in science. Veugelers et al., for instance, pitch their study in the context of a growing concern that public funding of research is

40 See Veugelers et al., 2022.
41 See Boudreau et al., 2016
42 Ibid., p. 2779. For a qualitative study on ERC practices, see Luukkonen, 2012. Note that Luukkonen too establishes a bias against what she calls "truly innovative" grant proposals, not against innovative proposals *per se*.
43 See Hofstra et al., 2020.
44 See Whitley et al., 2018.

> increasingly risk [/novelty] averse despite the fact that the science needed for breakthroughs is generally characterized as requiring risk-taking [...] Risk [/ novelty] aversion on the part of science funders may thus slow advancements in the knowledge frontier, impeding or delaying breakthroughs.[45]

The authors' findings then suggest that science funders do not function as they are supposed to: they *should* reward highly innovative proposals, but they do not. In other words, funders are less neomanic than they ought to be—or, conversely, conservatism is framed as a bad thing.

As salient in this context is the alarm prompted by a recent, already much-cited study indicating that articles and patents have become less disruptive (read: less highly novel) over the years—where an article's or patent's disruptiveness is the extent to which it changes the course of an entire field of study.[46] Conservatism is also condemned here. Most dramatically, it is said to pose

> substantial threats to economic growth, human health and wellbeing, and national security, along with global efforts to combat grand challenges such as climate change.[47]

In this account, (disruptive) scientific innovation is a panacea for all our ills; pleas for consolidation or slow science look almost criminal. And this brings us to neomania on the part of society.

Society

The argument I wish to develop here is a simple one. According to the first premise, society expects innovation to resolve its major challenges. According to the second, science is a key driver of innovation. From this, I conclude that society expects science to be strongly biased towards novelty.

Let's start with the first premise. The late Benoît Godin recounts how, over the centuries, innovation shifted from a vice to a virtue in the public perception.[48] For 2,500 years, the concept had negative connotations (I will say more about this in Chapter 3). Only over the

45 See Veugelers et al., 2022, p. 2.
46 See Park et al., 2023.
47 See ibid., p. 138.
48 See Godin, 2015.

course of the twentieth century, and especially since the 1960s, did it become tantamount to 'progress'. It became a catchword "that everyone understood spontaneously—or thought he understood—; that every theorist talked about; that every government espoused" (see Figure 2.2).[49] In fact, as Matthew Wisnioski argues in his recent *Every American an Innovator*, in the U.S. the injunction to innovate became a defining moral and civic ideal.[50] In Wisnioski's account, innovation moved from being a technical activity associated with post-war research and development to a universal obligation: a marker of good citizenship, personal authenticity, and national vitality. Beginning in the 1950s, the U.S. government and its industrial partners promoted innovation as both the guarantor of economic growth and the antithesis of bureaucratic stagnation. By the late twentieth century, Wisnioski argues, this rhetoric had seeped into schools, corporations, and cultural policy. 'Everyone an innovator' was not merely a slogan; it was a new model of selfhood, one in which to live creatively was to live rightly.

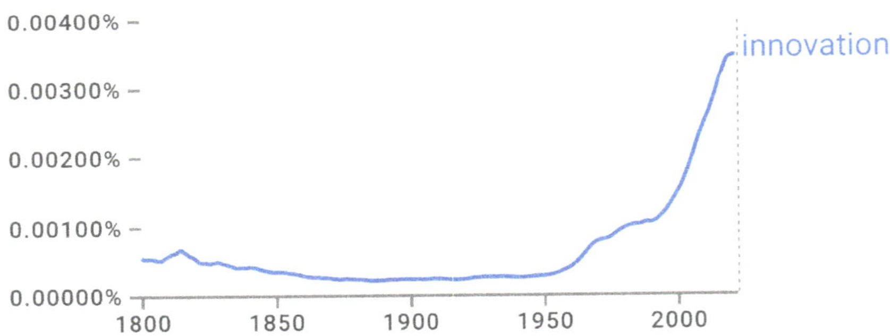

Fig. 2.2. Frequency of the term "innovation" over time (via Google Ngram). Note the steady increase since the 1960s.[51]

A telling example—provided by Godin[52]—of how innovation has become a panacea for every problem is a 2010 OECD document intended to contribute to the further development of developing countries, entitled "Innovation and the Development Agenda". The

49 See Godin, 2016a, p. 22.
50 See Wisnioski, 2025.
51 See also Chapter 3.
52 See Godin, 2015, p. 285.

OECD there champions a then new approach to development, one that is predicated on innovation: "[M]ost current social, economic, and environmental challenges require creative solutions based on innovation and technological advance".[53] Rather than starting from specific problems that specific developing countries face, the document treats innovation as the *a priori* solution. And the document disregards all that has proven effective in ending (extreme) poverty—non-innovative things like: peace, schooling, basic universal health care, electricity, clean water, toilets, contraceptives, and microcredits to get market forces started.[54] Further, as Godin rightly remarks, developing countries are primarily consumers of knowledge and technology produced elsewhere. Accordingly, their ability to further develop by absorbing what comes from outside deserves as much attention as their ability to further develop by their own inventive efforts. Indeed, some of the fastest growing countries in world history include Malaysia, Taiwan, Korea, and China—and their (initial) growth was primarily driven by imitation rather than invention.[55]

Another example concerns the United Nation's 17 Sustainable Development Goals (SDGs). Innovation itself is one of the SDGs (goal 9: "Industry, Innovation and Infrastructure"),[56] is mentioned in the descriptions of two others (goal 8: "Decent work and Economic Growth"[57] and goal 17: "Partnerships for the Goals"),[58] and is generally considered to be a means for achieving all the others.[59] So, it is not only consultancy firms talking in hyperboles about innovation ("Innovation Is the Only

53 See OECD, 2010, p. 30.

54 See Rosling, 2018.

55 See Edgerton, 2006.

56 Goal 9 reads: "Build resilient infrastructure, promote inclusive and sustainable industrialization and foster *innovation*" (see https://sdgs.un.org/goals/goal9, italics added).

57 Target 8.2 reads: "Achieve higher levels of economic productivity through diversification, technological upgrading and *innovation*, including through a focus on high-value added and labour-intensive sectors" (see https://sdgs.un.org/goals/goal8#targets_and_indicators, italics added). And target 8.3 reads: "Promote development-oriented policies that support productive activities, decent job creation, *entrepreneurship, creativity and innovation*" (ibid., italics added).

58 For instance, target 17.6 reads: "Enhance North-South, South-South and triangular regional and international cooperation on and access to science, technology and *innovation*" (see https://sdgs.un.org/goals/goal17#targets_and_indicators, italics added).

59 See Anadon et al., 2016, p. 9682.

Way to Win the SDG Race").[60] The UNDP commends innovation for its potential to eradicate poverty, protect the planet, build peaceful societies and prevent conflict, manage risks and improve disaster responses, and advance gender equality.[61] A similar brand of eco-modernism[62] has also long characterized Intergovernmental Panel on Climate Change reports. Indeed, it was only in 2022 that the Panel started to regard as viable an alternative to eco-modernism, to wit, degrowth—where degrowth relies more on political change than on innovation *per se*:

> Degrowth aims for intentional decreases in both GDP and coupled [greenhouse gas] emissions [...] using policy mechanisms such [as] a 'cap and share' framework for distributing emissions permits on an annually declining basis with legislation to prohibit the overshoot of established carbon budgets [...]. Degrowth thus seeks to minimise reliance on negative emissions technologies, such as the large-scale deployment of BECCS [Bioenergy with Carbon Capture and Storage] [...] and aims to generate progress toward achieving the SDGs by prioritising redistribution rather than GDP growth.[63]

Scientific research on innovation, which has been booming since the 1960s (and we will see why in Chapter 3), has a clear pro-innovation bias too. This bias means that in innovation science

> For the most part, researchers have asserted that *innovation is good*, and have sought the conditions under which innovation takes place, the conditions which facilitate its diffusion, and the characteristics which distinguish more innovative individuals, groups, and organizations from less innovative ones. One implicit assumption appears to be that *research should foster innovation* [...] Rarely if ever is not adopting an innovation considered to be a possibly important, adaptive strategy.[64]

This quote is from the 1980s, but little has been done to remedy the problem:[65] what largely remain left out of the scientific picture are such things as imitation, resistance to innovation, withdrawal (or ex-novation), slow innovation, adaptation, and maintenance, as well as

60 See https://www.bcg.com/publications/2019/innovation-win-sdg-race
61 See UNDP, 2018.
62 For a criticism of the SDGs along these lines, see the review in Kotzé et al., 2022.
63 See IPCC, 2022, p. 173.
64 See Kimberly, 1981, p. 88; quoted in Godin, 2015, pp. 283–284, italics added. The term "pro-innovation bias" comes from Rogers, 1976.
65 See Godin & Vinck, 2017.

the negative consequences of innovation (but see the recent scholarship described below).

In their book entitled *The Innovation Delusion*,[66] Lee Vinsel and Andrew L. Russell introduce the term innovation-speak, "a breathless dialect of word salad that trumpets the importance of innovation while turning that term into an overused buzzword".[67] Innovation-speak includes

> the word "innovation" and its variants (disruptive innovation, social innovation, impact innovation), along with a slew of Silicon Valley-approved jargon that business leaders and marketers have incorporated since the 1990s—change agent, thought leader, disruption, angel investor, intrapreneurship, design thinking, STEM education, unicorn, incubator, start-up, regional innovation hub, innovation ecosystem, innovation district—not to mention the habit of calling companies "lean" and "agile".[68]

Quite paradoxically, innovation-speak is not a discourse of optimism but of fear, the fear of being left behind—captured by the mantra 'innovate or die'—be it as a country or firm in the global economy, or as a student in the quest for a non-blue collar job. In Europe, the fear is often cashed out in terms of an 'innovation gap' with the U.S. and China that urgently needs to be closed, principally because the 'innovation gap' results in a 'productivity gap', and ultimately a 'GDP growth gap'. Europe, as the recent report by the former Chair of the European Commission Mario Draghi is at pains to stress, is tasked to innovate itself out of its deplorable status as a continent that is seriously drifting behind.[69]

More worrisome, innovation-speak (not innovation *per se*) harms society: it has led to a devaluation and neglect of maintenance and care. Indeed, maintenance isn't overly sexy; it is much easier to excite voters or capital investors by means of promises about shiny new things. Governments (and businesses, for that matter) equate success with growth, and growth with innovation. But, in the long run, maintenance of old things (water and sewer systems, public transport, roads,

---

66 See Vinsel & Russell, 2020.
67 Ibid., p. 10.
68 Ibid., p. 20.
69 See https://commission.europa.eu/document/download/97e481fd-2dc3-412d-be4c-f152a8232961_en?filename=The%20future%20of%20European%20competitiveness%20_%20A%20competitiveness%20strategy%20for%20Europe.pdf

power and communication systems, schools, hospitals, courthouses, production facilities) is as critical to growth as the production of new things. Vinsel and Russell amply illustrate how governments and industries, engrossed in innovation-speak, tend to forget this simple truth. Vinsel and Russell's is a story of collapsing bridges, dams and levees,[70] of companies raising profits at the expense of safety, of new educational technologies not delivering the goods they were supposed to deliver, of inequality in healthcare, of bubbles and of failures. It is also a story about innovation-speak perpetuating the 'maintainer caste': a caste that, despite carrying out critical work (and in contrast to the 'STEM caste'), falls at the bottom of status hierarchies.

Apparently, then, society thinks highly of innovation, due to its potential to cure many of our ills. We now need to establish the second premise, to wit, that science is commonly considered to be one of the primary drivers of innovation. For that, one just needs to look at existing scientific models of innovation: they virtually all attribute a key role to science in bringing about innovation. Let's briefly consider two of the most prominent models (I'll have much more to say about them in Chapter 3).

The simplest model is the so-called linear model, which can be represented as follows: basic research → applied research → development. Basic research, here, is research carried out for its own sake—that is, without a specific practical application in mind (e.g., quantum physics). Its results are subsequently used by the applied scientist, who generates scientific knowledge with a practical purport (e.g., research on semiconductors). Finally, in the development stage, what is learned from the two previous stages is channelled into new materials, devices, production systems and processes, etc. (e.g., the industrial production of LEDs). So, basically, the critical role that science plays is generating the necessary knowledge for producing new things. Although the linear model is typically presented as pertaining to technological innovation and marketable products, it need not be; the model may just as well represent the processes that lead from basic social science research

70 Vinsel and Russell's evidence comes primarily from the U.S., where public infrastructure notoriously suffers from neglect. You may find their account compelling, depending on where you work or live. It is compelling in Belgium (where I live) but less so in the Netherlands (where I work).

to the development of, say, new environmental policies, new ways of managing organizations, and new modes of public communication.

A currently much more popular model of innovation, called the triple helix model, also assigns an important role to science. Whereas the linear model assumes a linear process, with uni-directional interactions, the triple helix model acknowledges the non-linear nature of innovation, and the reciprocal nature of interactions. More particularly, according to the model, innovation is brought about by the mutual and close interactions of three actors (or helices): governments, universities, and industry. Further, according to the model, the interactions comprise more than, as is assumed by the linear model, flows of knowledge. For instance, universities do not merely supply industry with the scientific knowledge required to produce novel products; universities also provide funds, personnel, instruments, education, and the like, and such provisions often also go in the opposite direction (i.e., from industry to universities).

It is in this context that the aforementioned concept of the entrepreneurial university emerged. Rather than restricting itself to academic goals, the entrepreneurial university from the outset has in mind, much like the businesses they collaborate with, economic utility—or, as Etzkowitz puts it, "the 'capitalization of knowledge' is the heart of a new mission for the university".[71] Universities, thus, are supposed to serve economic development by galvanizing innovation.

To summarize the argument of this section: in a society that expects innovation to resolve its main challenges, and that treats science as a—if not *the*—main driver of innovation, one shouldn't be surprised to observe a scientific enterprise that distinctly loves the new.

Epilogue: Society's neomania, justified after all?

Our society, I have argued, suffers from neomania. But perhaps its preoccupation with innovation isn't pathological after all. Perhaps this preoccupation is justified because the net result of innovation (real innovation, not innovation-speak) is unequivocally positive. How can this be established?

71 See Etzkowitz, 2008, p. 27.

One way, popularized by Joseph Schumpeter, is to assess the economic impact of innovation. Schumpeter was one of the first to posit that innovation is a key driver of economic growth.[72] Hundreds of papers have been devoted to empirical verification of his doctrine, and the majority indeed find a positive correlation between innovation and increases in Gross Domestic Product (GDP).[73] So let us assume that the correlation is real, and that—and this is a non-trivial assumption—the arrow of causality goes from innovation to growth, rather than from growth to innovation (i.e., countries or firms spending their excess money on innovation). Would that be enough to justify neomania?

Arguably not. GDP is too blunt an instrument to capture all that is of value to us—think of the things that Vinsel and Russell are concerned about, such as non-collapsing bridges and equality in healthcare.

Obviously, I am not the first to question the adequacy of GDP as a measure of well-being. I could refer here to the degrowth movement, which rightfully points out that our economies' obsession with growth has led to widespread ecological damage, and which rightfully (or not; I am not sure) wonders how the cause of the problem—technological innovation—can be turned into a solution. But I don't want my assessment to be put away as too leftist. So, consider this.

In 2010, in the wake of the financial crisis, the renown economists Joseph Stiglitz (explicitly critical of the degrowth movement),[74] Amartya Sen, and Jean-Paul Fitoussi published a book, entitled *Mismeasuring Our Lives: Why GDP Doesn't Add Up*.[75] The study was commissioned by the then president of France, Nicolas Sarkozy (not your leftwing extremist, at least not according to European standards) who, in his foreword, voiced concern that current measures of economic performance, in particular GDP, are ill-equipped to tell us how well countries perform in terms of societal well-being and economic, environmental, and social sustainability. Indeed, Stiglitz and colleagues identify no less than eight

72 See Schumpeter, 1934.

73 See Cameron, 1998. Not all agree, though. Here is, for instance, the eminent economic historian David Egerton: "It was known in the 1960s that national rates of economic growth did not correlate positively with national investments in innovation, research and development, or innovation. It has not been the case that countries that innovate a lot, grow a lot" (Edgerton, 2006, p. 109).

74 See Stiglitz, 2019.

75 See Stiglitz et al., 2010.

dimensions of well-being that GDP measures miss out on: (i) material living standards (income, consumption, and wealth); (ii) health; (iii) education; (iv) personal activities, including work; (v) political voice and governance; (vi) social connections and relationships; (vii) environment (present and future conditions); and (viii) insecurity, of an economic as well as physical nature.

A thorough assessment of the net impact of innovation would thus be an assessment according to these eight dimensions. Unfortunately, for many of them we don't (yet) have any meaningful metric, let alone a compelling set of studies investigating the link between innovation and these alternative metrics. But let's see what we do have.

We do have plenty of studies assessing the relationship between innovation and unemployment. Interest in the relationship is not new—it has been with us minimally since Thomas Mortimer, in 1780, expressed the fear that machines would "exclude the labour of thousands of the human race, who are usefully employed".[76] Some 240 years later, with the rise of AI, Mortimer's fear is even more relevant—although it now pertains to white-collar jobs, not just blue-collar jobs.[77] However, in their recent systematic review (of 127 studies in total), Kerstin Hötte and colleagues reassure us that this fear is unwarranted: the majority of papers they examined document a positive rather than negative impact of technological innovation on employment.[78] Apparently, more often than not, new technologies create more jobs than they destroy. Perhaps this time, with the rise of AI, everything will be different, as some have cogently argued[79]—but to be honest, there remains much uncertainty.

Another important thing to come out of Hötte et al.'s study is more worrisome: low-skill, and increasingly middle-skill, workers are more likely to experience negative employment impacts of technological change than high-skill workers. Such skill-biased disappearance of jobs makes perfect sense: jobs increasingly involve the operation of increasingly sophisticated technologies, and jobs thus increasingly discriminate against those who lack extensive training.

Relatedly, there is evidence (primarily from the U.S.) that this also

---

76 See Mortimer, 1780, p. 145.

77 See, e.g., Brynjolfsson & McAfee, 2012; Ford, 2016; Sachs & Kotlikoff, 20122012.

78 See Hötte et al., 2023.

79 See Susskind, 2021.

applies to wages: innovation primarily puts pressure on the wages of low-skill workers, i.e., those whose jobs are the easiest to replace (by technologies, or—on account of the communication revolution—by offshoring).[80] Conversely, the wages of those who can keep up with the increased skill demands of sophisticated technologies go disproportionally up.[81] Further, digital technologies have created 'technological superstars': individuals whose earnings are orders of magnitude greater than those of everyone else, because they have been able to exploit their talent on a scale that was unthinkable before the information age (think: Silicon Valley unicorns).[82] So while innovation's net benefit might be positive, its fruits tend to be distributed unevenly—a point which even *Forbes* is willing to admit.[83]

These statistics say little about how people *experience* technological developments. Indeed, even if innovation were a net job-creator, it could increase, and plausibly has increased, rates of precarious employment—employment with not just low wages (see above) but also uncertain contracts. The rise of computerization and advances in communication technology have fuelled a trend towards foreign competition and the internalization of production. In order to rapidly respond to changes in a globalized market, organizations became more flexible—and this flexibility extends to hiring.[84] The rise of the platform economy (Uber, Deliveroo, and the like) has merely aggravated this.

Further, more speculatively, innovation may have led to an increase in what the late David Graeber called 'bullshit jobs'.[85] These are high-skill jobs that workers themselves find 'pointless, unnecessary, or pernicious' and that are primarily associated with 'information work' in the service economy. They include, among many others, HR consultants, communications coordinators, PR researchers, financial strategists, corporate lawyers, private equity CEOs, lobbyists, actuaries, telemarketers, bailiffs, and "the sort of people (very familiar in academic contexts) who spend their time staffing committees that discuss the

80 See Brynjolfsson & McAfee, 2012; Saint-Paul, 2008. See also the systematic review by Fragkandreas, 2022inequality.
81 See Kalleberg, 2011.
82 Ibid.
83 See Gaskell, 2019.
84 See Kalleberg, 2011.
85 See Graeber, 2019.

problem of unnecessary committees".[86] Although these jobs offer good compensation, their pointlessness undermines job satisfaction and well-being more generally.

Recent studies confirm that innovation may generally have a negative effect on subjective well-being. Aldieri and colleagues found negative correlations between countries' levels of innovation (as measured by the number of patents they produce) and these countries' happiness index data.[87] The latter are based on surveys asking respondents to rate their agreement with statements such as "I am satisfied with whatever I have", "I am happy in the career that I am in", and so forth. The negative effects were primarily mediated by inequality; innovation tends to increase income and health inequalities (see also above). A person's subjective well-being, now, is dependent on the characteristics of the people in their reference group. Inequalities that a person experiences between themselves and their reference group thus reduce the person's subjective well-being.

Also worth a careful read is a recent review by Biggi and Giuliani, entitled *The Noxious Consequences of Innovation: What Do We Know?* (published in the generally pro-innovation journal *Industry and Innovation*).[88] The authors remark, as I have, that scholarship in the field of innovation sciences predominantly lauds innovation as an undisputed engine of economic growth and that only recently have scholars started to pay attention to the harmful effects of innovation. Biggi and Giuliani's ambition, then, is to summarize the results of this new field of inquiry. Table 2.1 is an attempt at summarizing their summary. The authors distinguish between five major categories of noxious consequences: consequences pertaining to work (Theme 1, including self-neglect, burnout, and emotional exhaustion); to the unsustainability of technological transitions (Theme 2, including unsustainable energy consumption, environmental degradation, and poor labour conditions); to growth (Theme 3, including negative growth and inequality); to emerging technologies (Theme 4, including the possible harmful effects of nanotechnologies, big data, social media, etc.); and to open innovation (Theme 5, including negative effects on employment and wealth generation).

86 See Ibid., p. xiii.
87 See Aldieri et al., 2021, and references therein.
88 See Biggi & Giuliani, 2021.

Table 2.1. Summary of the results of Biggi & Giuliani (2020).

	Examples
Theme 1: Work-related consequences of technology acceptance.	Addiction to workplace technologies; workplace technologies causing work-family conflicts, self-neglect, and burnout; work-related social media use triggering emotional exhaustion.
Theme 2: Unsustainable transitions.	Excessive ICT use leading to unsustainable energy consumption; agricultural-improving technologies having unanticipated environmental degradation effects and causing poor farm worker labour conditions; 'rebound-effects' of innovations designed to contribute positively to the environment.
Theme 3: Innovation and growth downside effects.	Trade-induced technological changes having a negative impact on GDP and increasing pollution; financial innovations leading to exploitation of investors; innovation leading to inequality and skill-biased unemployment.
Theme 4: The dangers of emerging technologies.	Uncertainties regarding the possible harmful effects on public health and the environment of nanotechnologies, big data, the Internet of Things, cognitive computing, virtual technologies, and social media.
Theme 5: Open innovation's dark side.	Open innovation's negative effects on employment, firm performance, and wealth generation.

Surely, much more needs to be done to determine the exact relationships between innovation and well-being (in each of Stiglitz et al.'s eight domains). Yet, in light of the evidence that we *do* have (say, about environmental degradation, inequalities, threats to health and well-being) and the great number of unknowns (think of Stiglitz et al.'s list again), it would be a mistake simply to assume—as innovation scientists and policy makers tend to—that innovation is good full stop.

Conclusion

The preceding should not be read as a Luddite manifesto. I *am* a big fan of electricity; I do cherish my laundry machine; I admire the science that went into the life-saving drugs I and my cherished ones take. But there is a vast space between Luddism and the neomania I have documented above. It is this middle-ground that we need to reclaim.

Nor should the preceding be read as an argument that neomania alone explains the troubling state of science. While it plays a central role, other systemic factors—such as publish-or-perish pressures and publication bias—are equally important. Yet even this doesn't capture the full complexity. The causal structure is likely far messier, involving not just institutional incentives but also individual-level dynamics.

Researchers, like all humans, are prone to persistent cognitive biases: confirmation bias, anchoring bias (the tendency to rely too heavily on initial evidence), and observer bias can all subtly distort judgment and interpretation. In addition, many scientists may be intrinsically more drawn to novel, high-profile work than to the slower, less glamorous tasks of robustness-checking—the kind of work found in the RED AREA of Figure 1.2. This preference may reflect not only career incentives but also deeper psychological dispositions, such as curiosity, ambition, or a desire for recognition.

Beyond these individual and structural factors, cultural influences also shape scientific practice in ways that can interfere with truth-seeking. One important dimension is inter-cultural variation in epistemic norms—differences in what counts as legitimate evidence, acceptable methodology, or persuasive argumentation across national or disciplinary cultures. For example, some academic cultures place a strong emphasis on deference to senior scholars and established schools of thought, which can discourage younger researchers from openly challenging dominant paradigms or pursuing unconventional approaches. In such environments, intellectual hierarchy and tradition may take precedence over empirical scrutiny, slowing down the correction of flawed ideas.

Despite the importance of these psychological and cultural factors, the remainder of this book will focus primarily on the incentive structure of science. That focus is not meant to deny the relevance of other causes, but reflects a pragmatic choice: the incentive landscape is already complex enough, and it is here that reform seems both most urgent and most actionable.

But before we can begin to devise a cure, we need to understand the nature of the disease. Specifically, we must ask: where does our collective obsession with the new come from? What historical and institutional forces made the new so central to scientific value? This is the topic of the next chapter.

3. Neomania: A short history of the obsession

In what could be called—with an anachronism—a very early scientometric study, the historian Lynn Thorndike, in 1951, noted a remarkable pattern: the seventeenth century witnessed a dramatic increase in the use of words like "new" and "unheard-of" in the titles of scientific works.[1] Indeed, the seventeenth century produced numerous famous treatises with 'new' in the title, including, among many others, William Gilbert's *New Physiology* (1600), Johannes Kepler's *New Astronomy* (1609), Francis Bacon's *Novum Organum* (1620), Galileo's *Two New Sciences* (1638), and Robert Boyle's *New Experiments* (1660, 1665, 1682). But Thorndike in effect observes appeals to newness in virtually all spheres of seventeenth-century scientific activity: general methodology, astrology and astronomy, experimental physics, alchemy and anatomy, chemistry, mathematics, phytology, physiology, medicine, and engineering.

Where did this fairly sudden and pronounced predilection for novelty come from? The answer is that seventeenth-century natural philosophers intended to mark a clear, threefold break with the past. Let's see how this plays out in the writings of Francis Bacon (1561–1626), one of the most emblematic figures of the age of the new.

The first (and most well-known) break from tradition concerned methodology. Bacon's *Novum Organum* ("New Instrument") is a methodological treatise that meant to replace Aristotle's *Organon*—a collection of works about logic and the scientific method that at the time

1 See Thorndike, 1951; quoted in Godin, 2015. More generally, this chapter has benefitted enormously from the historical work of the late Benoît Godin (see references throughout).

 https://doi.org/10.11647/OBP.0507.03

enjoyed tremendous popularity. The *Novum Organum* famously deposes Aristotelian deductive reasoning as a fruitful approach to the study of nature, and develops a sophisticated form of inductive reasoning in its stead.[2] Rather than backing generalizations with one or two examples, or treating them as self-evident premises of a syllogistic argument, the modern scientist is supposed to inductively infer generalizations from careful observation and experimentation; rather than consulting books within the Aristotelian canon, the modern scientist must consult nature.[3] Or, as the historian Michael North puts it in his *Novelty: A History of the New*: "To be new and original, from Bacon onward, meant to disclaim ancient authority, and start over on one's own".[4]

A second break with the past concerns the subject matter of science. On this score, Bacon's programme both coheres with and departs from what was common at the time. It cohered with the popular idea that science ought to (re-)produce knowledge of nature as designed by its creator. This recovery of Edenic knowledge involved correcting the scientific errors accumulated over the centuries in the wake of Aristotle.[5] Yet, Bacon's programme also leaves room for genuinely *new* knowledge:

> There is another substantial reason why the sciences have made so little progress, namely, that it is not possible to keep a true and proper course if the goal itself has not been correctly placed and fixed there. Now the only true and proper goal of the sciences is to bring *new discoveries and powers to human life*.[6] (italics added)

The new discoveries and powers of Bacon's time were too significant to sweep under the Edenic carpet: they included the discovery of new celestial bodies and continents and the introduction of new inventions—Bacon was particularly fond of printing, gunpowder, and magnets, technologies that "changed the whole face and state of things throughout the world".[7] And he was hopeful that

---

2 Bacon's originality regarding induction is a matter of controversy (see, e.g., M. R. Cohen, 1926). No one denies, however, that Bacon's work captures the spirit of his day very well—and that suffices for my purposes. For a good introduction to Bacon's theory of induction, see J. Klein & Giglioni, 2020.

3 See Farrington, 1949.

4 See North, 2015, p. 48.

5 Ibid.

6 See Bacon, 1620, p. 81.

7 Ibid., p. 131.

> there are still many very useful things, which have not so far been discovered, hidden from us in the bosom of Nature, having no affinity or parallel with things already discovered, but quite off the beaten track of our imagination.[8]

("Off the beaten track"—Bacon would make for a great grant applicant!) What Bacon expresses here is what Hannah Arendt calls a *pathos* of novelty, "the almost violent insistence of nearly all the great authors, scientists, and philosophers since the seventeenth century that they saw things never seen before, thought thoughts never thought before".[9]

Bacon's third innovation—also illustrated by the indented quotes above, and closely related to the second innovation—is a radical redefinition of the goals of science. The scientist has a sacred duty to improve and transform the conditions of life, to support progress, by ridding the world of poverty, disease, and toil:[10]

> Natural philosophy shall not vanish in the fumes of subtle or sublime speculations, but shall be operative to relieve the inconveniences of man's estate.[11]

And

> arts and sciences ought to be like mines, loud on every hand with the sounds of new operations and further progress.[12]

Thus what is ordained by God is not, as the ancient Greek philosophers believed, moral speculation about good and evil, but rather the application of science to the "compassionate alleviation of the suffering of the human condition".[13] In the image of God, humans must create, not speculate.

Note that, for Bacon, science's usefulness is not yet, as it would become later, tied to commercial interests. For he despises people who, from the sum of human science and art, "take or seek only what they can turn to professional use or pecuniary advantage or enhanced

8 Ibid., p. 109.
9 See Arendt, 1958; quoted in Godin, 2015.
10 See Farrington, 1949.
11 See Bacon, 1624, p. 297.
12 See Bacon, 1620, p. 102.
13 See Mitcham, 1990.

reputation, or benefits of that kind."[14] (ibid.)

Bacon is not the only giant philosopher explicitly wedding science to progress. Here is René Descartes, plausibly influenced by his near contemporary (but without proper APA-style-referencing), regarding his own discoveries in natural philosophy:

> For by them I perceived it to be possible to arrive at knowledge highly useful in life, and in room of the Speculative Philosophy usually taught in the Schools, to discover a Practical, by means of which, knowing the force and action of fire, water, air, the stars, the heavens, and all the other bodies that surround us, as distinctly as we know the various crafts of our artizans, we might also apply them in the same way to all the uses to which they are adapted, and thus *render ourselves the lords and possessors of nature.*[15] (italics added)

Bacon's (and, more generally, the seventeenth century's) threefold novelty—an urge for methodological innovation, a craving for new knowledge, and an insistence on the link between science and societal progress—laid the foundations for contemporary science's neomania. Still, other things had to happen before neomania became the obsession it currently is.

Interlude: A note on the history of the concept of innovation

Despite the novelties he introduced, Bacon is careful enough not to call himself an innovator. For, at the time, innovation was still a subversive, political concept, one that meant introducing, in a revolutionary and violent fashion, change into the established order.[16] During the Reformation, the concept was used for discrediting people who hoped to modify religious doctrine. So, in 1548, Edward VI, King of England issued a *Proclamation Against Those That Doeth Innouate*: a warning intended to put off putative religious innovators.[17] During his Bacon's own life (1561–1626), the late Tudor and early Stuart eras were marked by dynastic uncertainty, the lingering effects of the Reformation, and

14 See Bacon, 1620, p. 90.
15 See Descartes, 1637, p. 108.
16 See Godin, 2016a.
17 Ibid. p. 24.

growing tensions between Parliament and the Crown. Innovation, in this context, was frequently associated with heresy, rebellion, and sedition.[18] In early Stuart England, even minor procedural changes in Parliament could be condemned as 'innovations', as illustrated by Sir Edward Coke's famous objection in 1624: "Innovations in Parliament [are] most dangerous".[19] In the seventeenth and eighteenth centuries, the term 'innovator' was also used by monarchs to discredit republicans.[20] And around the 1850s, it started to be used pejoratively to denigrate social reformers and early socialists. Their calls for radical change—whether in labour rights, property relations, or political representation—were seen by many conservatives as threats to the established social order.[21]

Still, over the nineteenth century, and in the Francophone world in particular, innovation started to get positive connotations as well, and its meaning widened to include science. For instance, in 1873, Alphonse de Candolle[22] publishes a ranking of countries according to the number of eminent scientists they produced, and tells us that English scientists are more *innovative* than their German counterparts.[23] In that same book, de Candolle examines the various social factors that favour scientific innovation, including religion, institutions, governments, language, funding, education, tradition, the size of the pool of scientists, and the proportion of mathematicians to natural scientists.[24] His work reflects an early attempt to quantify and sociologically analyse the conditions under which science flourishes—an approach that anticipates later scientometric and innovation studies.

And de Candolle is not alone in this: as I have documented elsewhere,[25] at the turn of the nineteenth to the twentieth century, virtually all French philosophers of science (e.g., Henri Poincaré, Pierre Boutroux, Ernest Naville, Élie Rabier, Pierre Duhem, Édouard Le Roy, Abel Rey, André Lalande, Théodule-Armand Ribot, Edmond Goblot, Léon Brunschvicg, and Jacques Picard) participated in a research programme devoted to

---

18 See Godin, 2016b.
19 See https://www.british-history.ac.uk/no-series/proceedings-1624-parl/apr-29
20 See Godin, 2016b.
21 Ibid.
22 See Candolle, 1873.
23 See Godin, 2015.
24 See Vaesen, 2021.
25 Ibid.

the logic, psychology, and sociology of scientific discovery. This loosely coordinated intellectual movement—sometimes referred to as the 'French neopositivist tradition'—sought to understand not just the products of science, but the processes and conditions that made discovery possible. In effect, these scholars were the first to take up some of the issues that later would be taken up by innovation scientists proper. For instance, Pierre Duhem, Pierre Boutroux, and Léon Brunschvicg carefully study the collective nature of discovery.[26] They challenge the idea of the lone genius and instead emphasize the role of shared conceptual frameworks, disciplinary norms, and cumulative knowledge.

Jacques Picard addresses the increasingly strong links between science and industry and laments the corresponding crisis in basic science. He calls for reforms in the recruitment of scientific personnel (to select personnel based on research rather than teaching qualities); science funding (to decentralize research and multiply the number of laboratories and to increase the overall governmental budget for science); education (to stimulate, following Dewey, experiential learning); rewards (to provide researchers with incentives to excel); international cooperation among scientists (to promote cooperation by protecting intellectual property rights); and scientific labour (to Taylorize[27] scientific production processes).[28] And so forth.

Innovation, in the nineteenth century, covers science but also the useful arts, including

> Agriculture, accounting, education, and the practical arts like warfare, mining, metallurgy, construction and textiles [...] [It includes] new artefacts (e.g., machines, artillery, fortresses, furniture), new methods (instruments and processes), new forms of organization (stables; farms) and new practices (education, medicine).[29]

The relationship between the useful arts and science is as Bacon conceived it: advances in the useful arts are facilitated by the application of (new) scientific methods, theories and principles. Our contemporary conception—of an intimate relationship between science, technology,

---

26 See Boutroux, 1914; Brunschvicg, 1912; Duhem, 1905, 1906.

27 Taylorization means breaking down jobs into simple, specialized, repetitive tasks and optimizing each step.

28 See Picard, 1928; see also, again, Vaesen, 2021.

29 See Godin, 2015.

and innovation—thus has its origins here.[30]

But in spite of its positive connotations, 'innovation' continued to be used also pejoratively. Nineteenth-century authors continue expressing the fear that innovation may inordinately upset tradition and received opinion; others warn of the societal risks posed by (technological) innovation.[31] This changes over the course of the twentieth century, especially after World War II. And it is governments—those that until then had most opposed innovation—who become innovation's greatest advocate. So what happened?

Firstly, in the wake of the work of Joseph Schumpeter, innovation became associated primarily with technology. Indeed, as we have seen, already prior to World War II, Schumpeter held that technological innovation is a key determinant of economic change. After the war, his ideas became *really* mainstream—in particular thanks to the work of Robert Solow.[32]

Solow aimed to resolve a mystery about the U.S. economy: traditional factors such as capital and labour couldn't explain the great bounty of the late 1950s.[33] He proposed that the residual growth was due to technological innovation. In his account, which almost instantly became orthodoxy, technological change increases productivity, which in turn supports economic growth.

Solow thus provided governments and international organizations (e.g., the OECD) with an additional lever in their attempts to spur economic progress. They soon developed and adopted 'innovation policies' (see below)—and these policies started really to proliferate when, in the late 1960s and early 1970s, economic growth appeared to decline. (In fact, low growth still is a principal motivation for policy makers to invest in innovation.)[34]

According to Matthew Wisnioski, what also happened (in the U.S.) is that from the 1950s onwards, political leaders and industrial planners cast innovation as the democratic antithesis of totalitarian stagnation.[35] Federal programmes such as the National Defense Education Act and

30 Ibid., p. 198.
31 Ibid.
32 See Solow, 1957.
33 See Vinsel & Russell, 2020.
34 Ibid., p. 28.
35 See Wisnioski, 2025.

the expansion of the National Science Foundation fused Cold-War competition with a belief in the moral productivity of creativity. By the 1970s, innovation had acquired a distinctly psychological dimension: it signified not only national security or economic progress, but self-realization. Management theorists, educators, and futurists promoted creativity training as a means of personal liberation. 'Innovative capacity' became a trait to be measured, cultivated, and displayed—an inner resource whose neglect was portrayed as civic failure.

During the 1980s and 90s this moralization of innovation intensified. The rise of Silicon Valley entrepreneurship, venture capital, and information-age rhetoric redefined technological novelty as a form of self-expression. As Wisnioski shows, the figure of the innovator merged with that of the entrepreneur: a risk-taking individual who simultaneously fulfilled market demand and personal destiny. Schools introduced 'innovation labs', cities founded 'innovation districts', and universities rebranded as 'innovation ecosystems'. To innovate was to belong—to participate in the continual reinvention of the self and the nation.

Along the way, as Godin observes in his *Innovation Contested: The Idea of Innovation over the Centuries*, the meaning of innovation shifted.[36] We saw that, at the end of the nineteenth century, innovation also referred to the scientification of the useful arts. In the first half of the twentieth century, innovation denoted the application of invention and science to industry (the scientification of industrial production). Finally, from the 1960s onwards, and as part of a larger trend to 'economize' science and technology, innovation was primarily associated with the commercialization of technical inventions.

Some things, however, did remain constant over the twentieth century (and up till now, for that matter). First, innovation was positively associated with novelty and originality. Second, science was thought to be a key driver of innovation.

I need to say a bit more about these two points. More specifically, for our story, we need to establish that, historically, science's strong orientation towards novelty—or, its neomania—in great part comes from its strong orientation to support innovation.

---

36 See Godin, 2015, p. 199.

The role of science in innovation

Rather than detailing the complicated history of twentieth-century theorizing about processes of innovation,[37] we will reconsider the two models of innovation from Chapter 2—viz., the linear and the triple helix model. We will not just see that according to each, science serves innovation (as I already argued in Chapter 2). We will also see that they have been highly influential: they have shaped science policy and accordingly, have pushed science (further) in the direction of neomania.

Recall that the linear model in fact refers to a class of models, the most basic of which can be represented as follows: basic research → applied research → development. The origins of the linear model may be traced back to the early twentieth century.[38] But, the model became engrained in governmental policies only after World War II: the U.S. National Science Foundation (NSF) adopted the model in 1953, the U.K. Department of Scientific and Industrial Research in 1958, and the OECD in 1962.[39] What 'adopting' means can be best illustrated with an example.

From its inception in 1950, the NSF's principal responsibility has been "to develop and encourage the pursuit of a national policy for the promotion of basic research and education in the sciences",[40] where basic research is

> The source of all technological progress [...] [It is] a continuing search for *new* knowledge [...] which is systematic, but without direction save that which the investigator himself gives it to meet the challenge of the unknown.[41] (italics added)

Applied research, then, is simply "the practical use of [this] newly found knowledge";[42] development, finally, "takes applied research out of the laboratory and translates it into production".[43] *Voilà*: the linear model.

This neat partitioning of innovation into stages allowed the linear

37 For a book-length account of that history, see Godin, 2017.
38 Ibid., pp. 61–76.
39 Ibid., p. 90.
40 See NSF, 1957, p. vii.
41 Ibid. p. 1.
42 Ibid., p. 1.
43 Ibid., pp. 1–2.

model to justify NSF's mission (and thus its existence). The narrative went more or less as follows. Science came relatively late to the New World. In the early days, 'Americans' were busy "[fighting] the Indians, [conquering] a continent of wilderness and [building] a Nation"; and, obviously, "a soldier or farmer does not produce higher mathematics".[44] Later colonists, if at all, looked to Europe for scientific ideas and education. Things began to change in the middle of the nineteenth century. The booming American industry and agriculture increased the demand for applied scientists and engineers—who were gradually trained in the newly established American universities. This trend became even more pronounced during and after World War I, and again during World War II, when the U.S. could no longer rely on Europe for its (intellectual) supplies.

> Forced to develop new laboratories and plants and processes, American industry began investing increasingly in *applied research* and *development*—and became more nearly self-sufficient in devising and providing the instruments and equipment it needed.[45] (italics added)

In other words, for nearly a century, American science had emphasized application and development. However, World War II—the atomic bomb in particular—had underlined the strategic importance of basic research. Post-war recovery would also crucially depend on such research. As Vannevar Bush, one of the principal architects of the NSF, put it in 1945:

> Progress in the war against disease depends upon a flow of new scientific knowledge. New products, new industries, and more jobs require continuous additions to knowledge of the laws of nature, and the application of that knowledge to practical purposes. Similarly, our defense against aggression demands new knowledge so that we can develop new and improved weapons. This essential, new knowledge can be obtained only through basic scientific research. [...] Advances in science when put to practical use mean more jobs, higher wages, shorter hours, more abundant crops, more leisure for recreation, for study, for learning how to live without the deadening drudgery which has been the burden of the common man for ages past. Advances in science will also bring higher standards of living, will lead to the prevention or cure of diseases, will promote conservation of our limited national resources,

44 Ibid., p. 10.
45 Ibid., p. 12.

> and will assure means of defense against aggression. But to achieve these objectives - to secure a high level of employment, to maintain a position of world leadership - the flow of new scientific knowledge must be both continuous and substantial.[46]

One would be hard-pressed to find a firmer statement of the idea that science is the great panacea for humanity's problems. In any case, the quote makes very explicit the link between science, novelty, and innovation: our insatiable thirst for "*new* products", "*new* industries", "*new* improved weapons", and so forth, can only be quenched by the "continuous and substantial" production of *new* basic scientific knowledge.

The U.S. government had traditionally spent considerable budgets on applied science (e.g., research carried out by the Naval Observatory, the Department of Agriculture, and industrial research laboratories); now it had to step up its game with respect to basic science—and the NSF became its main vehicle for doing that. The NSF still is the U.S.'s largest basic science funder, and it still is committed to supporting *new* basic research with the ultimate aim of fostering, as per its slogan, "Innovation anywhere, opportunity everywhere".

Importantly, grant-giving bodies across the globe have been modelled on the NSF—and thus have adopted the NSF's (implicit) agenda of novelty-seeking.[47]

Over time, linear models of innovation were complemented with non-linear models, the (still) most popular being the triple helix model. This model emerged in the 1990s and, as said, represents the non-linear, mutual interactions of three actors (governments, universities, and industry), whose aims and activities have come to substantively overlap:

> Universities, firms, and governments each "take the role of the other" in triple helix interactions even as they maintain their primary roles and distinct identities. The university takes the role of industry by stimulating the development of new firms from research, introducing "the capitalization of knowledge" as an academic goal. Firms develop training to ever higher levels and share knowledge through joint ventures, acting a bit like universities. Governments act as public venture capitalists while continuing their regulatory activities. In contrast to

46 See Bush, 1945, p. ix.
47 See Kealey, 2023, and for the Dutch case Kersten, 1996.

> theories that emphasize the role of government or firms in innovation, the triple helix focuses on the university as a source of entrepreneurship and technology as well as critical inquiry.[48]

Rather than being tasked with the conservation of knowledge (as in the Middle Ages), or with the generation of fundamental-but-potentially-useful-knowledge (as in the linear model), the contemporary—that is, entrepreneurial—university is from the outset oriented towards economic development:

> The university's assumption of an entrepreneurial role is the latest step in the evolution of a medieval institution from its original purpose of conservation of knowledge to the extension and capitalization of knowledge. As the university increasingly provides the basis for economic development through the generation of social and intellectual, as well as human, capital, it becomes a core institution in society.[49]

Capitalization of knowledge here means that the entrepreneurial university has in mind and exploits the commercial potential of the knowledge it produces, through patenting and licensing, stockholding, consultancy work, technology transfer, incubating new firms, and taking the lead in regional renewal efforts.

The triple helix is both an empirical and, for our purposes more importantly, normative model. So, on the one hand, the model captures the triple helix interactions that have emerged bottom-up (first, in the 1980s, in the U.S.—later elsewhere). On the other, it provides innovation policy makers with a blueprint for how to organize and spur (regional) innovation. For instance in Europe, from the later 1990s onwards, triple helix interactions were promoted 'top-down' in response to the innovation gap between the U.S. and Europe.[50] Exemplars of successful triple helix innovation clusters include Silicon Valley (U.S.), Silicon Fen (Cambridge, U.K.), BioValley (Germany-France-Switzerland), Medicon Valley (Denmark-Sweden), Brainport (Eindhoven, the Netherlands), and the Oslo Innovation District (Norway). Countries with a strong tradition of implementing the triple helix model include the countries above as well as, among many others: Japan, South Korea, Singapore,

48 See Etzkowitz, 2008, p. 1.
49 See Etzkowitz, 2002, p. 1.
50 See Etzkowitz, 2003, p. 109.

Israel, Finland, Spain, Taiwan, Australia, India, Chile, Brazil, and Canada. Accordingly, the entrepreneurial university has *de facto* become the ideal of many a modern university.[51]

Why and whence the triple helix?

In the 1980s in the U.S., and in the 1990s in Europe, the number of patent filings by universities increased dramatically (see Fig. 3.1). One of the likely factors responsible for this explosion was the passage, in 1980, of the Bayh-Dole Act in the U.S., and the adoption of Bayh-Dole-inspired legislation across Europe in the 1990s.[52]

The Bayh-Dole Act established a legal framework that encourages universities (and other recipients of public research money) to derive patents from their publicly funded research. The idea behind the Act is that giving patent ownership to universities (rather than to their governmental sponsors), and thus giving universities the opportunity to generate patent-derived income, incentivizes them to innovate—that is, to develop new products and industries.[53] This was deemed necessary because of increased international competition and because of the industrial downturn of the 1970s.[54] The latter, so one thought, was in part the result of firms being unwilling to take the risk of further developing government-owned ideas. If a firm were to use such ideas for the development of a marketable product, its competitors would be entitled to produce the same product—for the underlying ideas would be funded by public tax money. No firm, thus, would be motivated to invest in the new product.[55] To resolve this free-rider problem, the Bayh-Dole Act encouraged universities to take up roles that previously only firms could take up; it broke down the clear barriers that existed between universities and industry (as championed by the linear model of innovation) and paved the way for an intensification of interactions

51 Ibid., p. 112.

52 See Sterckx, 2010; see also Radder, 2019. Other factors include the general boom in biotechnological research, and the *Diamond v. Chakrabarty* ruling, which considerably expanded the types of items that were up for patenting (e.g., micro-organisms became patentable).

53 See Rai & Sampat, 2012.

54 See Etzkowitz, 2002, p. 118.

55 Ibid. p. 113.

between the two (of the sort championed by the triple helix model).

Governments did something else to spur these developments: they made the funding of universities dependent not just on scientific output, but also on efforts to capitalize their knowledge.[56] No wonder that nowadays any self-respecting university (in the U.S., Europe, and elsewhere) has a Technology Transfer Office (TTO)—an office that supports researchers in protecting intellectual property rights, negotiating licensing agreements with industrial partners, and forming spin-off companies. As an intermediary between academia and industry, TTOs are a key feature of the Era of the Triple Helix (a terrific title for a Netflix series, for that matter).

Finally, governments offered tax credits or deductions to businesses that set up research collaborations with academic institutes, and grants to academic institutes that set up research collaborations with businesses. They sometimes served as venture capitalists and helped to set up angel investor networks. Governments thus came to have several levers to galvanize innovation. Whereas in the linear model their role was restricted to funding basic research, they now intervened at various levels, by means of an entire suite of legal and financial instruments.

Going off the dramatic increase in university-owned patents over time (see Figure 3.1), these instruments seem to work—not least because by now they have been translated into local policy: as said, funding agencies and universities customarily rely on patenting and licensing as an evaluation criterion in their grant, hiring, and promotion decisions.[57] The link with science's neomania should be obvious. Patent laws stipulate that patents are granted only for inventions that meet the criteria: non-obviousness (the invention is not an obvious combination or variation of existing technology to someone with ordinary skill in that field), utility... and novelty. Novelty, according to the European Patent Office, means that "there must be no evidence that [the] novel aspect of your idea has ever been described before, or used for the same purpose before".[58] The novelty requirement thus prevents one from patenting something that in actuality belongs to the public domain.

---

56 Ibid.

57 Not soon enough, according to some: see Sanberg et al., 2014.

58 See https://www.epo.org/en/learning/learning-resources-profile/business-and-ip-managers/inventors-handbook/novelty-and-prior-art#:~:text=For%20your%20idea%20to%20be,for%20the%20same%20purpose%20before

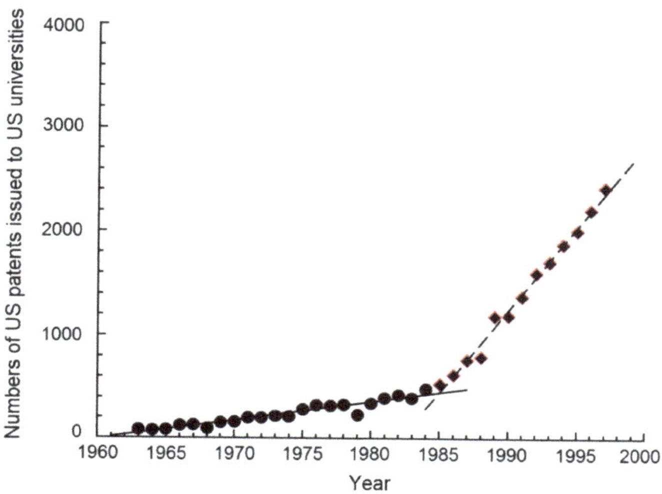

Fig. 3.1. Number of U.S. patents issued to U.S. universities over time. Note the steep increase around the early 1980s—an increase following the Bayh-Dole Act. Circles are data points for the years 1963–1984, squares are data points for the years 1985–1997. Figure adapted from Tseng & Raudensky (2014).

Interestingly, the provisions of the Bayh-Dole Act may have contributed to the proliferation of low-quality patents, in two senses. First, academic patents' average plausible utility has gone down. In an early study, Henderson and colleagues observed that the dramatic growth rate of academic patenting hardly matched the relatively small expansion of university spending on research and development.[59] Accordingly, "either universities had uncovered great economies in their operation, or else more junk was getting patented".[60] Given the relatively low number of citations for the patents in question, Henderson and colleagues conclude the second.

Second, and more relevant to our overall story, 'low-quality' may refer to the science underlying patents: the rush to be the first invites researchers to file a patent before the science that went into it is settled.[61] It is in fact nothing more than an instantiation of the link Merton posited between the priority rule and questionable research practices—but now applied to patents. A recent study by Janet Freilich, entitled *The*

59 See Henderson et al., 1998.
60 See Mirowski, 2011, p. 306.
61 Ibid., p. 309.

Replicability Crisis in Patent Law, substantiates this.[62]

Patents often contain experimental data. Such data serve a threefold purpose: substantiating that the invention works; proving that the patent filer in effect invented the technology; and describing how to produce or deploy the invention to putative patent users. Freilich then examined 500 biomedical patents and assessed the methodological quality of the reported experiments. Her findings are sobering:

> Only 62% of experiments in patents in my sample disclosed sample size, 12% were randomized, 4% were blinded, 2% conducted replicate studies, and 63% had statistical analysis of any kind.[63]

These results are worse than the results of studies investigating the methodological quality of scientific articles. Working under the reasonable assumption that methodological quality is a valid proxy for replicability, Freilich concludes that patent experiments are irreplicable at rates at least as high as experiments in scientific journals. This conclusion holds irrespective of who is filing the patent, academics or industry. Apparently, the cost of filing a patent is not high enough to deter inventors from submitting non-working technologies—suggesting that some pursue a shotgun strategy, filing broadly in the hope that at least one idea will prove viable or profitable.

We have seen at least one factor that might justify paying a patent fee for non-working technology: it helps to boost (or game, some would say) one's university ranking. For such rankings, and Reuter's innovative university ranking in particular, take into account such factors as 'patent volume' (i.e., number of patent applications) and 'patent success' (i.e., number of *successful* patent applications). As said, high ranks come with many benefits—in terms of recruitment of students and staff, and in terms of extra investments. Paying a fee for non-working technology may thus be rewarded in the long run.

What about the social sciences and the humanities?

One might think that the pressure to innovate and, in particular, the pressure to capitalize on knowledge are only felt in disciplines that

62 See Freilich, 2020.
63 Ibid., p. 434.

generate knowledge that is easy to commodify—e.g., the biological, chemical, physical, engineering, biomedical, and pharmaceutical sciences. Indeed, research in psychology, sociology, and economics, as well as research in the humanities, typically is not eligible for patent protection. But these disciplines too are strongly incentivized to partake in innovation, if not to contribute to the attainment of economic aims.

To start with, the linear model can be, and in fact has been, applied to the social sciences. When, in 1958, the NSF started to fund basic social science research (including research in the humanities, particularly the history, philosophy, and sociology of science) it hoped that, ultimately, the new knowledge generated by such research would lead to commercially interesting applications, and benefit "specialists concerned with such matters as management, public relations, communications, economic policy, public opinion, and political propaganda".[64] The NSF's rhetoric hasn't changed much since then. At present, its support of the social sciences is motivated by their contribution to "empower[ing] America's industries and communities, provid[ing] opportunities for entrepreneurs and creat[ing] the knowledge that fuels American economic leadership".[65]

In fact, as one of the first bodies to fund basic social science research, the NSF set a trend as to how to do so. To counter attacks from natural scientists who had their doubts about the scientific status of the social sciences, as well as attacks from conservative politicians who complained that the social sciences tended to serve left-wing ideologies, the NSF only funded approaches to social science that met universal scientific standards, such as generalizability, verifiability, and objectivity.[66] In other words, social science funding was, and still is, modelled on the mould of natural science funding—even if by now the NSF has expanded its set of evaluation criteria to include, e.g., novelty and impact.[67] So much of what pushes natural scientists towards innovation pushes social scientists in the same direction. True, agencies may have become more inclusive when it comes to the type of innovation they envisage, mentioning non-economic alongside economic impact. The

64 Ibid., p. 13.
65 See https://new.nsf.gov/sbe
66 See Solovey, 2015; Vaesen & Katzav, 2019.
67 See https://www.nsf.gov/pubs/policydocs/pappg22_1/pappg_3.jsp#IIIA

U.K.'s Economic and Social Research Council, for instance, states that their support of social science "helps raise productivity, address climate change, improve public services and generate a prosperous, inclusive, healthy and secure society".[68] Still, the council's predilection for novelty[69] is motivated by a now familiar agenda of (even if not necessarily technological) innovation.

Given its insistence on the importance of university-industry interactions, the triple helix model does seem to exclude social science and humanities research (SSH) from its area of application. Yet, again in funding schemes modelled on the triple helix, SSH research is increasingly called upon to buttress processes of innovation, even of *technological* innovation. As an example, consider the European Commission's Framework Programmes for Research and Innovation (FPs). Before the eighth FP, which was called Horizon 2020 and which ran from 2014 to 2020, FPs supported SSH research in two ways: there were special SSH funds, dedicated to research on the main social, economic, cultural, and political challenges of Europe; and there were schemes aimed at promoting the integration of SSH with innovation-oriented STEM research. Horizon 2020, however, skipped the first and funnelled virtually all of its SSH budget into the innovation-oriented second stream.[70]

Even worse, the role assigned to SSH in that stream was merely a supportive one. Horizon 2020 was framed as a programme that contributed to realizing the European Union's ambition to become an 'innovation union', and as such to secure its global competitiveness.[71] To achieve this, according to the then European Commissioner for Research, Innovation and Science, the idea was to 'embed' SSH across all of the Union's innovation goals: SSH researchers would (need to) make sure that "the social, political and human aspects are not forgotten alongside the technological aspects".[72] And because, according to the Commissioner, "[STEM] researchers, businesses, innovators and

68 See https://www.ukri.org/who-we-are/esrc/who-we-are/, Who ESRC is – UKRI

69 See https://www.ukri.org/councils/esrc/guidance-for-applicants/impact-innovation-and-interdisciplinarity-expectations/

70 See Keraudren, 2018.

71 See Felt, 2015.

72 See https://ec.europa.eu/commission/presscorner/api/files/document/print/en/speech_13_740/SPEECH_13_740_EN.pdf, p. 3

entrepreneurs are of course the primary customers of Horizon 2020", SSH scholars were treated as an add-on rather than as equal partners.[73]

This is not just a matter of unfortunate framing. In its execution also, Horizon 2020 put SSH research at the service of technological innovation. The European Commission even admitted this much in its own intermediate evaluation of Horizon 2020: "By [integrating the SSH], H2020 pays tribute to the findings that, although research in technologies can provide technical solutions to major challenges, Social Sciences and Humanities (SSH) can help making them accepted, understood and appropriated by the general public".[74] The social sciences and humanities were thus seen merely as tools to achieve public acceptance of innovations brought forth by the natural and engineering sciences.[75] In other words, as subordinates, they helped to capitalize on the knowledge produced elsewhere.

Discourse analyses of policy documents and funding calls suggest that, at least on paper, the current FP—called Horizon Europe—might better live up to its promise of SSH integration and its promise to address societal (not just market-oriented, technological) challenges, including adaptation to climate change, the development of climate-neutral cities, and the restoration of oceans and bodies of water.[76] The intended link between SSH and industry (and thus commercial interests) remains close, though. According to the Horizon Europe programme guide, SSH researchers are principally tasked to design and guide processes of so-called social innovation—that is, innovation that, e.g., through processes of co-creation, engages citizens from the outset.[77] Social innovation

> concerns the development of new products, methods, and services for and with society involving citizens, public authorities, business and industry, and academia — the Quadruple Helix — in their design, development, and implementation.[78]

SSH researchers thus remain instrumental to the development of new

---

73 Ibid., p. 2.
74 See https://op.europa.eu/en/publication-detail/-/publication/4365f75a-5efe-11e8-ab9c-01aa75ed71a1/language-en; quoted in Stamm, 2019, p. 390.
75 See Kropp, 2021; Kropp & Larsen, 2023.
76 See Puputti, 2022.
77 See https://ec.europa.eu/info/funding-tenders/opportunities/docs/2021-2027/horizon/guidance/programme-guide_horizon_en.pdf, p. 22.
78 Ibid.

products, methods, and services. What has changed is their role: they are charged with *integrating* the fourth helix—viz., citizens—with the three helices of the triple helix—viz., government, academia, and *industry*. So, like earlier FPs, Horizon Europe expects the social sciences and the humanities to work closely with commercial partners with one single aim: innovation.

Conclusion

From the seventeenth century onward, science has embraced novelty. The embrace, we have seen, has only tightened. Over time, the scientific enterprise was turned into a genuine enterprise: an undertaking that came to serve innovation, which itself came to serve (economic) progress. This, then, is the sense in which neomania in society drives neomania in science.

This chapter concludes Part I of the book. According to our diagnosis, science is in a state of poly-crisis—one with multiple, overlapping failures—and a key culprit is our (society's) deep-seated neomania. In Part II, we will look at possible cures. We start by looking at a necessary but insufficient cure: open science.

PART II
THE CURE

4. Open science: Necessary but not enough

'Never waste a good crisis!', Brian Nosek, one of the founding fathers of the *Reproducibility Project: Psychology* must have thought. Around the same time the results of the *Reproducibility* project came out, Nosek and some of his colleagues published two articles, entitled *Scientific Utopia I* and *Scientific Utopia II*, both describing avenues to a future where truth prevails and knowledge in effect accumulates.[1] The papers have been called the most comprehensive discussion of possible reforms aimed at resolving the reproducibility crisis.[2]

And indeed, many subsequent proposals are variations on themes developed by—or at least nicely written down by—Nosek and colleagues. Many of their ideas (some of which were definitely already around at the time of their writing), and the ideas they provoked, go under the banner of open science; and they are intended to deal both with A-crises and B-crises. As Nosek and colleagues write in *Scientific Utopia II*,

> Opening our research process will make us feel accountable to do our best to get it right [i.e., to increase the quality of published findings, in other words, to deal with what we have called A-crises; KV] and, if we do not get it right, to increase the opportunities for others to detect the problems and correct them [i.e., to increase the number or researchers engaging in the robustness-increasing activities of what we have called the Red area, in other words, to deal with B-crises; KV]. Openness is not needed because we are untrustworthy; it is needed because we are human.[3]

1 See Nosek et al., 2012; Nosek & Bar-Anan, 2012.
2 See Bausell, 2021.
3 See Nosek et al., 2012, p. 626.

https://doi.org/10.11647/OBP.0507.04

In this chapter, I will first give an overview of the many senses in which science could/should become open, and highlight why open science indeed contributes to resolving science's poly-crisis. Subsequently, I argue that, while necessary, open science is not enough. In particular when it comes to B-crises, opening up science—or more generally, removing barriers blocking the RED AREA in Figure 1.2—will not suffice, and this for two reasons. First, there are various obstacles to opening science—in other words, there are barriers to removing the barriers. Second, removing barriers to the RED AREA will not automatically lead to RED AREA work; we need things that actively *tempt* people into the area. That is, we need to compel researchers to actually start *re-using* the vast amounts of data that are (already) openly shared.

What's this thing called open science?

Open science may mean a number of things.[4] In its most basic sense, it refers to sharing data and analytic files to improve the reproducibility of research. So Nosek et al., in their *Scientific Utopia II*, tell us that open science is the ultimate solution to many of the problems that science currently faces, and that open science impels researchers to be open with regard to their data, methods and tools, and workflows. Regarding the first, open data enables "confirmation, extension, critique, and improvement of research already conducted".[5] Second, open materials and tools have the same effect but, additionally, facilitate replication. Indeed, only studies that contain a detailed description of their study design and methodology allow others to re-run them. Opening workflow, finally, means being transparent about the entire research process—e.g., about possible conflicts of (financial) interests; about the ways in which the results of a study might be affected by personal *a priori* beliefs, as well as confirmation and other biases; and about whether the findings of a study were predicted or rather a serendipitous discovery. This second-order evidence is useful to assess the quality of results and, again, it

4 See Banks et al., 2019.
5 Ibid., p. 624.

facilitates replication. Note that each of the three practices—sharing data, sharing materials and tools, sharing workflow—target both A- and B-crises: they increase the quality of findings, and facilitate further development of those findings.

A second sense of open science pertains to the ways in which science communication is organized. *Scientific Utopia I* proposes six changes aimed at improving such communication: (1) fully embracing digital communication, and completely abandoning paper copies; (2) removing pay-walls and going fully open access; (3) publishing prior to review; (4) creating a system of peer review that works independently from journals; (5) publishing referee reports alongside articles; (6) solicit referee reports from the entire scientific community and let the review process go on indefinitely. Some of these measures primarily target A-crises, others are geared towards resolving B-crises, and still others do both.

(1) Abandoning printed journals has two principal advantages. Printed journals have a limited amount of space. Researchers thus cannot communicate everything that is essential for others to replicate their study. Going digital removes this barrier (a solution to B-crises; although some digital journals still have unforgiving word limits). Further, paper journals typically have a long lag between the submission of a manuscript and its eventual appearance in print. In an attempt not to get scooped by others, researchers thus are incentivized to rush their research even more than when their paper gets published online almost immediately upon acceptance (as in, e.g., *PLoS* journals) or even prior to peer review (see point (3)). Accordingly, fully embracing digital communication likely increases the quality of studies that are published (a solution to A-crises).

(2) Pay-walls are an obvious barrier to replication (researchers have no access to the original paper, nor to the results of other replications) and, more broadly, to the other activities in the RED AREA. Removing this barrier—i.e., by going fully open access—therefore contributes to resolving B-crises.

(3) Publishing prior to peer review, for example on pre-print repositories like *arXiv*, targets both types of crises. When it comes to A-crises, pre-printing helps to early establish the priority of a

discovery, (again) reducing the need to rush research. Further, pre-printing substantially reduces publication bias, as no one can bar publication of a negative result. Pre-printing thus may relieve some of the pressure exerted on researchers to engage in questionable practices that increase the likelihood of obtaining a positive result—e.g., p-hacking (the practice of manipulating data or statistical analyses until non-significant results become statistically significant), HARKing (Hypothesizing After the Results are Known: the practice of presenting a post hoc hypothesis—formulated after analysing the data—as if it were an *a priori* hypothesis). Additionally, pre-prints are live documents which authors can freely adjust and revise, for instance to correct honest mistakes. Regarding B-crises, pre-printing accelerates dissemination so that others can start their replications (and other robustness-increasing activities) early, and strike while the iron is still hot.

(4) *Scientific Utopia I* proposes that researchers submit their manuscript not to a journal, but to a dedicated review service, which does not decide whether the manuscript is fit for any particular journal but rather gives the manuscript a general grade. And because all articles are published (as per point (3)), journals are not really publishing papers, but rather promoting or endorsing them (e.g., all articles on a particular topic with a grade of A or higher).[6] A major benefit of this change is that it would dramatically shorten the review cycle: a paper typically gets reviewed once, not various times—that is, each time the paper is submitted to a different journal until it gets accepted somewhere. Shortening the review cycle, again, reduces the need to rush research (solution to A-crises). *Scientific Utopia I* furthermore proposes that the review service deploys generic evaluation criteria—such as soundness of the research, but not its discipline- or journal-specific estimated impact. This means that negative studies and replication studies also have a fair chance

6 As Thomas Reydon—one of the generous reviewers of this book manuscript—pointed out, this proposal potentially leads to the disappearance of journals altogether. Indeed, it's uncertain what role journals would still be able to play. Given their entrenched power, such a shift is unlikely to occur anytime soon—making Nosek's proposals seem more utopian than realistically achievable.

of getting a high grade and thus of getting endorsed by a journal (solution to A- and B-crises).

(5) There are two ways in which publishing non-anonymized referee reports alongside reviewed articles likely improves the quality of published findings (solution to A-crises).[7] First, reviewers would feel accountable to do their best to write high-quality reports, and such high-quality reports improve the quality of reviewed work. Indeed, good reports point out confounding factors, alternative analysis strategies, new avenues of investigation, and alternative explanations for reported data; and they constructively suggest strategies for addressing the concerns they have raised. Second, open, non-anonymous reviews allow researchers to earn credit for their refereeing work. Nosek and Bar-Anan even suggest that a new type of scientist may emerge—one who contributes to scientific progress not (only) by doing their own research, but by detecting deficiencies in the research of others. In fact, such scientists already do exist. The Dutch microbiologist Elisabeth Bik spends a considerable amount of her time on detecting photo manipulation in scientific publications, and has uncovered over 4,000 potential cases of misconduct—and is widely renowned for that.[8] The Belgian professor in building physics Hans Janssen devotes 20% of his time to assessing the merit of scientific articles[9] and reporting shoddy research practices to the editors of the journals these articles are published in. For example, he regularly skims through the most recent issues of the journal *Construction and Building Materials* and, when he does, he typically manages to quickly find a faulty paper, with flaws ranging from miscalculations, errors in analytic model derivation, and outright fraud. In his "The reviewer is

7 Nosek and Bar-Anan allow for the possibility that authors choose to opt out of non-anonymity. On a different note, reviewer Thomas Reydon pointed out several limitations of this proposal. Fewer researchers might be willing to serve as reviewers—an issue already pressing, given the current difficulty in finding suitable candidates. Moreover, junior reviewers could fear retaliation from senior researchers whose work they critique. Accordingly, the proposal may lead to lower rather than higher quality research, and thus risk exacerbating rather than resolving the very crises it aims to address.

8 See https://www.timeshighereducation.com/news/science-detective-mission-stamp-out-shoddy-research

9 Personal communication.

not always right" weblog, he details his experiences with editors and authors, and how they deal with the issues that he raises (often but not always as they should).[10] The cancer researcher Jennifer Byrne became a literature watchdog after she detected various errors in papers pertaining to DNA research. Michael Dougherty is philosophy's plagiarism detective. And the number of (part-time) error sleuths seems to be on the rise.[11]

(6) Making the review process open—such that anyone, not just a couple of hand-picked, solicited reviewers, can partake in reviewing—will increase the quality of published findings (solution to A-crises). Most of the journals issued by the European Geoscience Union (EGU), including, e.g., *Earth System Dynamics* and *Climate of the Past*, have such open peer reviews. Any paper which passes a basic quality check is put online and made available for public review and discussion. Designated reviewers and all interested parties may provide comments, and authors get the opportunity to respond. Additionally, the open review procedure could be made to continue indefinitely. Such continuous crowd-sourcing practices are more likely to identify, for example, critical confounding factors (and thus contribute to the resolution of A-crises). Again, in response, authors can revisit and improve their work; they can also do so when, for instance, the results of a replication conflict with the original findings.

In addition to the practices covered in *Scientific Utopia I* and *Scientific Utopia II*, the Center for open science adds registering reports and preregistration. The first refers to conducting peer review even prior to data collection. Authors thus submit for review a document describing their study's hypothesis, methods, and analysis plans. If the proposal gets accepted, they then follow through with the registered protocol, with the guarantee that the results will be published, be they positive or negative. This format is designed to reward responsible research conduct and, conversely, to eliminate Questionable Research Practices such as low statistical power, selective reporting of results, data manipulation, and

10 https://www.researchgate.net/publication/369795194_THE_REVIEWER_IS_NOT_ALWAYS_RIGHT

11 See https://retractionwatch.com/2018/06/17/meet-the-scientific-sleuths-ten-whove-had-an-impact-on-the-scientific-literature/

so forth—the format thus primarily addresses A-crises. Preregistration purports to do the same but rather than submitting their protocol for review, authors, in advance of their study, submit their research plan to an open repository—a digital platform, such as the Open Science Framework (OSF), that provides free, immediate, and permanent access to research materials.

Barriers to removing the barriers

To the extent that they address B-crises—whether they actually do remains an open empirical question (see also f.n. 4 and f.n. 5)—open science practices are of a removing-the-barriers kind: they make it easier for researchers to engage in replication studies and other RED AREA work, but do not *compel* them to do so. Sharing data, materials, tools, and workflows facilitate replication and so forth, but do not incentivize researchers *per se* to actually reuse shared data, materials, tools, and workflows. The same holds true for practices such as removing journal space limits by going digital, open access publishing, pre-printing, and dedicated review services. Before I explain, in the next section, why this is problematic, I here consider various factors that impede open science.

The first is that, since data may be hard to collect, authors have good reasons to be possessive over their own data.[12] Indeed, in the current hyper-competitive scientific regime, data often form a great competitive advantage in the research community; and in light of this, authors may be disinclined to give away the fruits of their hard work.[13]

Such possessiveness has been termed the 'Gollum effect' (after the greedy character from Tolkien's *Lord of the Rings* trilogy), and a recent survey of over 400 biologists (broadly conceived) across 45 countries suggests that the effect is widespread (at least among biologists, broadly conceived).[14] Over 40% of respondents indicated they had experienced

12 See Banks et al., 2019.
13 Ibid.
14 See Valdez et al., 2024.

the effect, and two thirds of them stated that it had substantively impacted their career—including prompting them to leave academia altogether. 20% of respondents even admitted to potentially having exhibited Gollum-like behaviours themselves.

It is easy to condemn such practices, but in a sense they are understandable—and perhaps even excusable. As long as the hyper-competitive nature of current science remains unaddressed, open science—in the sense of sharing data, materials, tools, and workflows—is unlikely to fully deliver.

Relatedly, data sharing standards are primarily set by rich institutions, such that they match these institutions' material capabilities and theoretical perspectives.[15] This seriously complicates the life of researchers who work in low-resourced environments (e.g., researchers in low-income countries). For one, these researchers often lack the reliable, powerful connectivity and computing resources to either share their data or reuse the data of others. For another, their theoretical approaches may very well conflict with the approaches championed by rich institutions. The latter could for instance favour de-contextualized data (e.g., genomic sequences of crops), whereas researchers in low-resourced environments may favour local information (e.g., data on the socio-cultural context of crops). Accordingly, power imbalances limit which types of data in effect get shared.

The increasing reliance on privacy-sensitive data forms another obvious obstacle to open science—and this goes beyond the restrictions that privacy laws (e.g., GDPR, HIPPA, FERPA) place on data handling. As Gabriel and Wessel point out, requiring raw data submission may be problematic for three key reasons.[16] First, organizations will be less likely to sign up for studies that involve data pertaining to them—among other things, organizations do not want to run the risk of data hacks on servers that are out of their control. Second, sensitive populations will be reluctant to consent to participating in studies. Given that accidental disclosure of their identity may have serious ramifications,

15 See Leonelli, 2023, p. 21.
16 See Gabriel & Wessel, 2013.

marginalized groups (e.g., people with mental health diagnoses or invisible physical illnesses, religious or political minorities) have good reason to stay clear from serving as study subjects. Third, raw data sharing may affect the types of variables one can meaningfully assess; responses based on sensitive variables—e.g., self-reporting of unethical behaviour or (under-)performance, health-related variables—may get seriously distorted.

Relatedly, application-oriented research may be unable to meet open science demands due to intellectual property interests.[17] This is a problem in particular in the current science regime, where the activities of businesses and academic institutions often are deeply intertwined (see Chapters 2 and 3), and both businesses and academic institutions consider their data a source of strategic business advantage.[18] Without changes to the regime, science will probably for a great part remain closed.

And then there are many individual-level barriers inhibiting researchers from sharing data openly. Based on the extant literature, Zuiderwijk and colleagues identified over 100 such factors, which they put in eleven categories: "the researcher's background" (e.g., older researchers are less inclined to share data); "requirements and formal obligations" (e.g., lack of funder requirements to make data accessible); "personal drivers and intrinsic motivations" (e.g., fear that data will not be reused); "facilitating conditions" (e.g., lack of required data-storage infrastructure); "trust" (e.g, fear of misinterpretation of shared data); "expected performance" (e.g., reduction in productivity as a result of data sharing); "social influence and affiliation" (e.g., research culture focused on publications rather than data); "effort" (e.g., time investments in data management); "the researcher's experience and skills" (e.g., lack of data management skills); "legislation and regulation" (e.g., difficulties in understanding licenses); and "data characteristics" (e.g., lack of clear data standards).[19] In fairness, Zuiderwijk et al. also identified many factors that *drive* data sharing (again, over 100), so there is no reason for despair. Still, even if drivers and inhibitors were to cancel

17 See Guzzo et al., 2022.
18 See Jones & Dages, 2013.
19 See Zuiderwijk et al., 2020.

each other out—assuming there is a way that they could do that—there is the issue of the *reuse* of open data. And that issue arguably limits the scope of open science in resolving the crises even more than the barriers to data sharing.

The under-usage of open research data

Let's start the section with another of Zuiderwijk et al.'s findings. The authors also identified drivers and inhibitors of the *re-usage* of open research data. Again, the factors are classified in eleven categories (the same as the categories mentioned above). And again, given the sheer number and wide variety of factors listed (over 100 each), it is impossible to tell whether drivers trump inhibitors or vice versa—unless one looks at the literature investigating the actual reuse of research data. That admittedly still small body of literature suggests that open research data are rarely reused, and thus that, plausibly, inhibitors tend to outweigh drivers.

Navar et al. (2016) studied the data-reuse of clinical trial data published on three prominent open access platforms, and found that of the more than 3,000 datasets only 15% had been requested by a limited number of other researchers.[20] In another study, Quarati and Raffaghelli (2022) analysed the usage data of over 6,000,000 datasets in the open data repository Figshare—a platform that classifies resources according to twenty-one scientific disciplines, ranging from Biology and Engineering, to Social Science and Humanities.[21] The usage data Quarati and Raffaghelli looked at were pretty basic: number of views and number of downloads. For both variables, distributions of frequencies are heavy-tailed, with very few datasets having a high frequency of use and most of them having zero reuse. By way of illustration, Figure 4.1 gives the distributions for the 'download' variable. Note that downloading, let alone viewing a resource—particularly when it concerns a figure or journal paper—does not necessarily mean that the downloader will actually engage in RED AREA activities (replication, generalizability checks, theory development, etc.). Therefore, by implication, Figure 4.1 is likely an *over*-estimation of actual RED AREA work.

20 See Navar et al., 2016.

21 See Quarati & Raffaghelli, 2022. For a broader discussion of data under-use, see Santos-Hermosa et al., 2023.

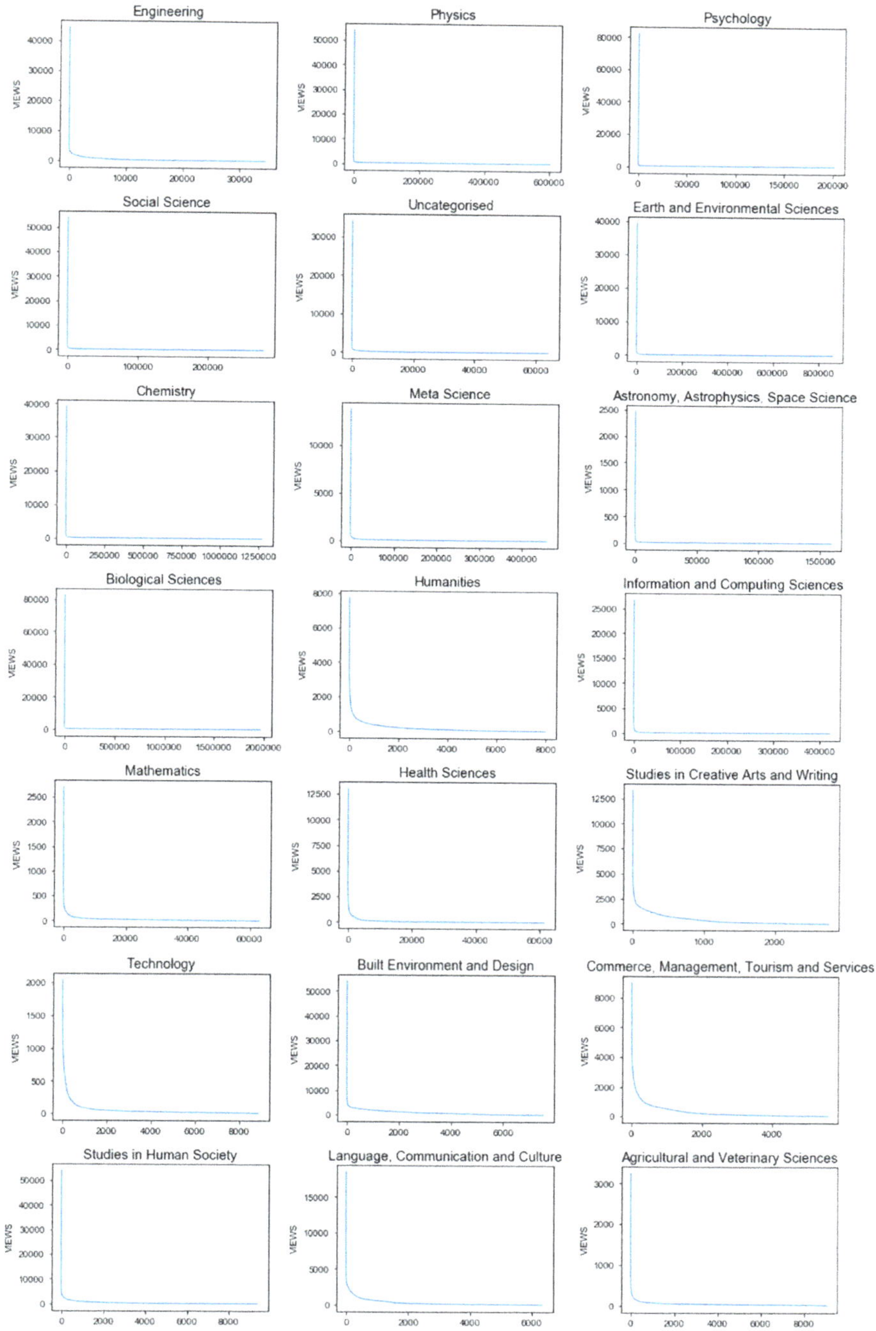

Fig. 4.1. Download distributions for various disciplines. X-Axis: dataset identifier; Y-Axis: number of downloads. Quarati & Raffaghelli (2022).

Interestingly, Quarati and Raffaghelli also explored the relationship between open data quality and reuse—hypothesizing that high metadata (i.e., data describing the main data) quality would prompt higher reuse rates. However, the authors found no evidence that the use of Figshare resources was associated with good metadata publishing practices.

If low quality is not a barrier to reuse, something else must be. As mentioned, Zuiderwijk et al. give us quite some individual-level factors that impede reuse. But a plausible, and more structural explanation is neomania: under-usage is to be expected when researchers are strongly pushed to pursue their own idiosyncratic research questions, and need their own idiosyncratic data to answer them. In fact, studies suggest that even researchers who reuse data tend to prioritize novelty over reproducibility. In a study of the National Heart, Lung, and Blood Institute data repository, 72% of data reuse was aimed at generating new insights and informing the design of new research projects—while only 4% focused on re-analysing primary outcome findings.[22] In any case, as long as novelty is over-rewarded, open science will partly miss its target: researchers might produce lots and lots of open data, but do so for a barely existent audience.

This is not a minor issue: although estimates are hard to come by, it is certain that billions are spent yearly on open science. Between 2015 and 2018—still in the early days of open science—scientists paid more than a $1 billion in open access publishing fees to the five biggest commercial scientific publishers alone.[23] Researchers funded by grants from the U.S. National Science Foundation, National Institute of Health, and Department of Energy are estimated to spend, on average, nearly 6% of their grant money on research data management and sharing (DMS) (including staffing, technology, and infrastructure, which cover the numerous activities of Table 4.1).[24] Applying this percentage to the estimated total budget that is yearly allocated by U.S. federal funding agencies, namely about $36 billion,[25] the total yearly amount needed to cover data sharing costs can be estimated at $2.1 billion in the U.S. alone—and this is a gross under-estimation, because the number doesn't include money spent by scientists who are funded by non-competitive

22 See Coady et al., 2017. See also Waithira et al., 2024.
23 See Butler et al., 2023.
24 See Hofelich Mohr et al., 2024.
25 See Vaesen & Katzav, 2017.

direct government money. And, given that the requirements for DMS are getting ever more strict (e.g., there is a move towards requiring shared data to be FAIR, i.e., findable, accessible, interoperable and reusable), DMS costs and time investments will rise globally—for instance, health researchers in the Netherlands are advised, as a rule of thumb, to reserve 5–10% of their research budgets for FAIR data stewardship.

On a final note, there are no signs that the open science movement is taking up the issue of under-usage. For example, reuse is not mentioned in the Center for Open Science's mission statement, according to which it envisions "a future scholarly community in which the process, content, and outcomes of research are openly accessible by default".[26] Nor is reuse part of UNESCO's recommendations for open science,[27] nor the European Commission's construal of it.[28] Nor is it a practice that any of these actors think needs to be incentivized. For example, one of the noteworthy incentives that the Center of Open Science and the European Commission propose is the Open Science Badge. Researchers may be awarded such badges when they adhere to open science principles (and will get recognition for them, e.g., in hiring and promotion). But the principles in question only pertain to sharing data and materials, and preregistration—there is no badge for reuse.

Table 4.1. Activities relating to data management and sharing (DMS).

Data Plan	Drafting a DMS plan at the proposal stage in compliance with relevant requirements, getting the plan reviewed and accepted by relevant assessment committees (e.g., ethical review boards).
Data Storage	DMS plan management, including: data repository management, curating the data, metadata management, data ingest, data security, and privacy.

26 See https://www.cos.io/

27 Here is UNESCO's definition of open science: "an inclusive construct that combines various movements and practices aiming to make multilingual scientific knowledge openly available, accessible and reusable for everyone, to increase scientific collaborations and sharing of information for the benefits of science and society, and to open the processes of scientific knowledge creation, evaluation and communication to societal actors beyond the traditional scientific community". See https://www.unesco.org/en/legal-affairs/recommendation-open-science.

28 See https://research-and-innovation.ec.europa.eu/strategy/strategy-research-and-innovation/our-digital-future/open-science_en

Data Integrity	Activities that include: licensing, archiving, preserving, infrastructure upgrade, and other data maintenance tasks that ensure data integrity and data availability to the research community.
Monitor	Includes ongoing monitoring, oversight, and audit activities to ensure the approved DMS plan is being complied with.
Post Closeout	Includes maintaining the data after the period of performance is completed: both administratively and financially (e.g., determining how data storage and retention will be paid for).

Conclusion

The open science movement has achieved a lot. In no more than a decade, researchers have become accustomed to practices that were earlier unheard of, including open access publishing, data sharing, preprinting, and preregistration. Quite a few journals have taken to heart the movement's suggestions to open up peer review, assess submissions on methodology alone, and run assessments post-publication; and open science has become a crucial pillar in the workings and policies of highly influential international organizations such as UNESCO and the European Commission.

But there is a 'but'. While open science practices have the potential to help address science's poly-crisis, most conspicuously by removing the barriers to replications and other follow-up work, they are not sufficient to attract researchers to take up such work—currently, open data appear not to get reused enough. To unlock the full potential of open science, we need something extra. And that something is coordination: researchers have to coordinate their work with one another, such that RED AREA work effectively gets done.

5. Coordinating our way out of the poly-crisis

Even in an ideal world where open science successfully removes barriers to data reuse, it would still only be part of the solution. Additional measures would still be needed to actively compel researchers to reuse open data and engage in RED AREA activities.[1] The current chapter argues that coordination—in particular, what I will call Peirce-style coordination—can help us to achieve this objective; its implementation is the subject of the next chapter.

But before we start, let's get clear on what researchers need to be coordinated into.

Slow is beautiful

Give time to research programmes

One way of thinking about the issue is that science needs to slow down—the development of ideas costs time. To convince you that this is a sensible idea, it may help to look at what prominent theories about the growth of scientific knowledge have to say.[2] They too emphasize that, for knowledge to grow, a continuous turnover of new ideas is a bad idea. Science, they propose, should strike a delicate balance between novelty-seeking and conservatism, between creativity and scepticism.

Let's first consider Popper's picture of scientific progress. In 1959,

1 RED AREA activities are the robustness-increasing activities presented in the red rectangle of Figure 1.2. They comprise reliability and internal validity assessment, external validity assessment, and theory development.

2 For a similar treatment, see Cohen, 2017.

 https://doi.org/10.11647/OBP.0507.05

Popper likens the process of the growth of scientific knowledge to the Darwinian process of natural selection:

> How and why do we accept one theory in preference to others? [...] We choose the theory which best holds its own in competition with other theories; the one which, by natural selection, proves itself the fittest to survive. This will be the one which not only has hitherto stood up to the severest tests, but the one which is also testable in the most rigorous way.[3]

For species to 'grow'—i.e., to accumulate beneficial modifications over time—through natural selection, three conditions must be fulfilled. The first is that there must be variation among individual organisms (variation). The second is that modifications are preserved in successive generations (inheritance). And the third is that individual organisms compete in a struggle for existence (selection).

For scientific knowledge to 'grow'—i.e., for it to accumulate beneficial modifications over time—one also needs variation, inheritance, and selection. Inheritance here refers to the process of transmitting bits of knowledge from one generation of scientists to another; variation is brought about by scientists formulating novel conjectures; and selection involves strictly testing these novel conjectures.

Note the importance of selection. Consider the case where there is inheritance and variation but no selection. Modifications would still accumulate, indeed, but they would not necessarily be beneficial. The stock of scientific beliefs would drift randomly (according to the rate of variation) rather than growing towards, say, higher verisimilitude. This, then, is the problem with neomania in science: so little attention is paid to selection that random drift becomes a genuine risk.

Note also what selection amounts to according to Popper: it comprises *repeated* testing of conjectures. This follows from his insistence on falsification. Conjectures can only be proven wrong. Accordingly, one failed attempt at falsifying doesn't say much; only when a conjecture survives multiple severe falsification attempts do we have reason to provisionally accept it. Repeated testing, however, has no place in a science that is organized according to neomanic principles. Put differently, for science to progress—rather than merely to produce a flood of untested conjectures—a sufficient number of resources should be spent on falsifying new conjectures.

3 See Popper, 1959, p. 108.

Despite its many flaws,[4] Popper's evolutionary account of scientific knowledge has been—and continues to be—highly influential.[5] Another influential account is Thomas Kuhn's. And, again, this account doesn't presuppose anything that resembles an obsession with novelty—although the title of the book it was published in, viz., *The Structure of Scientific Revolutions*,[6] may suggest otherwise.

Kuhn argues that science progresses in a cyclical fashion (see Figure 5.1). Regimes of "normal science" are followed by regimes of "revolutionary science", which are subsequently followed by regimes of "normal science" again (and so forth). "Normal science" refers to scientists working within and further developing the dominant scientific paradigm, where a paradigm is a whole way of viewing and examining the world. During a period of revolutionary science, the dominant paradigm is abandoned and replaced with a new one. These shifts are radical, and happen "when nature has somehow violated the paradigm-induced expectations that govern normal science".[7] Once scientists have settled into the new paradigm, they start working according to its dictates, resulting in a new episode of normal science.

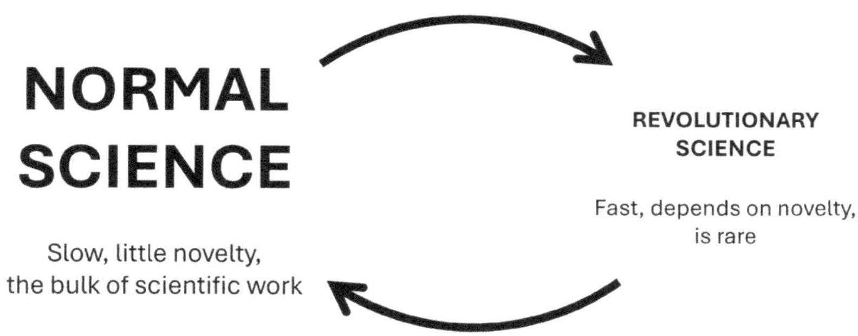

Fig. 5.1. According to Kuhn, science progresses in a cyclical fashion. Normal science (slow, little novelty) does and should form the bulk of scientific work. Image by the author.

4 For one, whereas biological evolution is blind (variation occurs randomly, and selection depends on environmental fitness), scientific change is intentional (scientists purposefully design experiments, revise theories, and argue rationally). For another, Popper's model of variation and selection arguably overemphasizes competition among theories, thereby downplaying the cooperative and cumulative dimensions of scientific inquiry (Maxwell, 1972).

5 For an overview, see Nanay, 2011.

6 See Kuhn, 1962.

7 Ibid., p. 52.

All this might suggest that novelty seeking in actuality *is* the focus of the *Structure of Scientific Revolutions*. But the book is just as much a book about normal science as it is about revolutionary science. And normal science doesn't tolerate novelty:

> No part of the aim of normal science is to call forth new sorts of phenomena; indeed those that will not fit the box are often not seen at all. Nor do scientists normally aim to invent new theories, and they are often intolerant of those invented by others. Instead, normal-scientific research is directed to the articulation of those phenomena and theories that the paradigm already supplies.[8]

Further, this kind of normal-scientific work forms the bulk of what scientists ordinarily do:

> Few people who are not actually practitioners of a mature science realize how much mop-up work of this sort a paradigm leaves to be done or quite how fascinating such work can prove in the execution. And these points need to be understood. Mopping-up operations are what engage most scientists throughout their careers.[9]

Normal science is, and should be, conservative, according to Kuhn. Whereas for Popper a single falsification suffices to give up on a theory, Kuhn recommends that scientists stick to their guns and—to avoid chaos, and for reasons of efficiency—only abandon a paradigm if it exhibits a critical mass of anomalies. Until then, "the results [of normal-scientific work] are significant because they add to the scope and precision with which the paradigm can be applied".[10] If anything novel were to come out of normal-scientific work, it would be of an incremental rather than 'ground-breaking' kind.

Concerning the rate of scientific change, Kuhn distinguishes between periods of accelerated change—viz., revolutionary science—and periods of relative stasis—viz., normal science. Science will not progress if it is in a permanent state of revolutionary science: to genuinely bear fruit, revolutionary ideas need to be developed by normal-scientific work:

> The areas investigated by normal science are, of course, minuscule; the enterprise now under discussion has drastically restricted vision. But

8 Ibid., p. 24.
9 Ibid.
10 Ibid., p. 36

> those restrictions, born from confidence in a paradigm, turn out to be essential to the development of science. By focusing attention upon a small range of relatively esoteric problems, the paradigm forces scientists to investigate some part of nature in a detail and depth that would otherwise be unimaginable.[11]

So, in Kuhn's account, once seeded, novel ideas should be given the time to grow. In other words, there is substantive merit in the dictum 'slow is beautiful'.[12]

Allow time to engage the public

In her *Another Science is Possible: A Manifesto for Slow Science*,[13] the Belgian philosopher Isabelle Stengers argues that Kuhnian normal science is not (yet) slow (enough).

Stengers notes that normal science operates within the strict confines of the paradigm and, for reasons of efficacy and efficiency, should be blind to extra-paradigmatic questions (raised by, e.g., policy makers or the general public):

> Kuhn [..] emphasis[es] [...] the need for the autonomy of research questions with regard to their social value or interest. For Kuhn, the paradigm is what determines the right questions. What comes from the outside [e.g., society] and cannot be identified as a puzzle would only perturb the course of normal, cumulative, science. So paradigmatic sciences have to be protected from the expectations they may kindle in society.[14]

And Kuhn's ideal, Stengers contends, corresponds to the ideal of fast academic science shaped during the nineteenth century,

> a model of research that promoted as a general ideal the *fast*, cumulative advance of disciplinary knowledge along with a correlative *disregard for any question that would slow this advance down*.[15] (italics added)

11 See Kuhn, 1962, p. 24.

12 A similar argument may be developed based on other prominent theories of the growth of knowledge, including Hull's (see, e.g., Cohen, 2017) and Lakatos' (1970)—but, for reasons of clarity, I will restrict myself here to Popper and Kuhn.

13 See Stengers, 2018.

14 Ibid., p. 95.

15 Ibid., p. 98.

Stengers' manifesto, then, is a plea for admitting 'decelerating' questions into scientific debates.[16] What are those?

They are certainly not questions raised by industry. For, also since the nineteenth century, science and industry have been natural allies:

> The dimensions that correspond to fast science's categories are rather naturally the very ones that are relevant for industrial development, since both agree on ignoring the same type of complications.[17]

Decelerating questions, rather, are ones that force scientists to take a pause and think about how "their concern for 'facts' [could] include the way these facts come to matter for other collectives".[18] That is, scientists ought to be given time to question the objectives of their research agendas, and to try to put their agendas into accord with those of relevant societal interest groups.

The following two examples will help to clarify Stengers' point. First, in the biomedical sciences, there appears to be a serious mismatch between the questions addressed by researchers and the questions that patients and clinicians want to see addressed.[19] For instance, patients with osteoarthritis of the knee favour research into physiotherapy, surgery, and educational and coping strategies. Despite this, researchers tend to predominantly focus their research on the development of drugs[20]—not coincidentally the area where industrial stakes are highest. Furthermore, researchers' methods are often ill-equipped to address patients' concerns: while patients with rheumatoid arthritis tend to be most concerned about fatigue, researchers typically assess the efficacy of treatments in terms of pain.[21]

Decelerating, in this example, would mean engaging patients in the development of research questions and designs from the start, e.g.,

16 The term "decelerating questions" comes from Conway, 2018.

17 See Stengers, 2018, p. 117. Stengers' picture is a bit muddied here. On the one hand, she believes that practicing scientists tend to be insulated from extra-paradigmatic concerns; on the other, she seems to think that practicing scientists are in actuality not insulated from the dictates of industry. So I read her as saying that to the extent that scientists currently consider extra-paradigmatic questions at all, these questions come from industry.

18 Ibid., p. 84.

19 See Chalmers & Glasziou, 2009.

20 See Tallon et al., 2000.

21 See Hewlett et al., 2006.

by means of community-engaged participatory research.[22] Although doing so might cause delay, it makes the research system far more effective: researchers would address genuine problems rather than problems conceived from the ivory tower or, perhaps even worse, by pharmaceutical marketeers.[23]

A second example concerns the mismatch in agricultural research between what scientists have focused on since the Green Revolution—viz., the search for high-yielding crop varieties—and the wishes of the supposed primary beneficiaries of such crops—viz., smallholder farmers. The latter would benefit more from community-based, collaborative research than from research based on controlled field trials. But, as Daniel Sarewitz notes, such community-based, collaborative research "is not the type of science that attracts major philanthropic or government funding; nor does it yield the rapidly publishable results that scientists need for professional advancement".[24]

How would deceleration in the sense of Stengers contribute to the resolution of the crises sketched in Chapter 1? Because international trade agreements don't allow direct subsidies to industry, governments seem to support industry in a roundabout way; they sponsor universities to serve as cheap research and development departments for industry. Such public-private partnerships should no longer be the golden standard of public engagement. Deceleration then implies that, when setting a research agenda and choosing research methods, scientists no longer by default turn to industrial partners—partners that, relative to scientists, have an even bigger stake in getting positive, novel results as quickly as possible. Decoupling the pace of industrial innovation from the pace of scientific development relieves quite a bit of the pressure to publish rapidly and, by the same token, frees up time necessary for responsible research practices—and, indeed, such practices *do* require additional time and effort.[25]

Further, Stengers' deceleration directly addresses the applicability crisis, which, as you may recall, pertains to the fact that scientific theories

---

22 See Wallerstein et al., 2018.
23 For a successful example of this, see Wit et al., 2014. See also Chapter 6.
24 See Sarewitz, 2016.
25 See Hostler, 2023.

tend to have low relevance to people's everyday lives.[26] It should be obvious why this is so: end-user engagement is built by design into research practices that are modelled according to Stengers' suggestions.[27]

Accepting that science is slow, or thinking bigger

Slow science, we have seen, emphasizes the importance of careful scrutiny of ideas (think again about the activities in the RED AREA); and, as per Stengers, of engaging non-scientific interest groups (not necessarily industrial ones) in setting research agendas and methods.

In practice, this means that we need to get used to thinking in longer timescales (again?).[28] Given the complexity of the problems it addresses, science usually progresses slowly. Even seemingly simple questions may take years and years to get settled. For instance, what are the short-term health effects of dietary salt reduction? In 1972, the consensus was that salt induces, among other things, hypertension. Four decades and 167 randomized controlled trials later, however, the evidence suggests that there is no clear connection between salt reduction and short-term health benefits.[29]

This is sobering news, since problems that require a quick resolution—such as the climate crisis—are far more complex than the relationship between salt and health. The scientific consensus about the anthropogenic nature of climate change took several decades to develop, and that was arguably the easiest part of the puzzle; no semblance of a consensus remains when it comes to the costs of and best pathways for preventing or addressing future impacts.[30]

There are at least two mutually non-exclusive responses to this. The first is to bite the bullet and accept that difficult questions can only be

26 See Berkman & Wilson, 2021.

27 Stengers says little about how deceleration would work in research areas with minimal practical purport. What are the relevant societal interest groups of, say, particle physics or of philosophy? Perhaps the relevant community here is just the community of practicing particle physicists or philosophers. So rather than individual researchers developing their own idiosyncratic agenda, there would be a community-based agenda, set after wide consultation of community members. For an example of such a practice in particle physics, see Garisto, 2023.

28 See Frith, 2020.

29 See Graudal et al., 2011; Sarewitz, 2016.

30 Ibid.

addressed in a slow fashion—and often cannot even be answered by scientific means alone. Indeed, many of today's societal challenges are of, what Alvin Weinberg called, a 'trans-scientific' kind. Trans-scientific questions are questions asked of science, yet "which cannot be answered by science".[31] Sarewitz provides an instructive example:

> Are genetically engineered crops necessary for feeding a burgeoning global population? [...] Are genetically modified crops more productive than conventional crops? Some researchers prefer to answer this question by looking at field trials that allow variables like weather and soil type to be carefully controlled. Others prefer surveys of actual farms, because they reflect real-world variability. These two approaches often yield contradictory results, and there is no way of adjudicating based on the data which of the two provides a better guide to the future.[32]

Or consider a question we encountered in Chapter 2: 'Which strategies are best to address the climate crisis, eco-modernist or degrowth strategies?' Although epistemically speaking this question looks answerable by science, it *transcends* science; data incompleteness, the complexity of the problem, and the implication of non-epistemic values (e.g., political and moral values) all prevent us from answering the question in a purely scientific fashion. Put differently, how to respond to climate change is as much a political as a scientific question. Scientists have a big responsibility here. The general public increasingly expects science to solve societal problems.[33] Surely, these problems need to be high on scientists' agendas. But scientists also have to exercise what Weinberg calls 'selfless honesty': they must publicly identify the limits of scientific facts—i.e., they must make explicit at which point scientific questions turn into political ones.

Getting our expectations about science right is one option, thinking bigger is another. One thing the Covid crisis has taught us is that, when scientists decide to head in the same direction, some parts of science can effectively proceed incredibly fast. Aligning scientists and encouraging them to consider the bigger picture may accelerate the resolution of scientific questions that, under a regime of slow science, might take too long to answer. And this requires a particular form of coordination.

---

31 See Weinberg, 1972, p. 209. See also Sarewitz, 2016.
32 See Sarewitz, 2016.
33 See Evans, 2014.

Coordination, Peirce-style

To see what type of coordination is called for, let's contrast it with what is not. According to an early, still influential account of scientific coordination, due to the chemist-philosopher Michael Polanyi, the main organizational principle of science is—and should be—the *self-coordination* of individual scientists by mutual adjustment.[34] This may sound quite abstract, so let's illustrate it with one of Polanyi's own favourite examples, viz., the solution to a large and complex jigsaw puzzle. Supposing that we engage a legion of helpers, how best to organize this process? What would not work is to hand out to each helper a number of randomly-selected pieces of the puzzle and let them work separately on their lot, because few of the pieces given to any helper would be found to fit together. We need some way of coordinating the individual actions of the helpers. The best we can do, according to Polanyi, is to let the helpers work on putting the puzzle together in sight of the others, so that

> every time a piece of it is fitted in by one helper, all the others will immediately watch out for the next step that becomes possible in consequence. Under this system, each helper will act on his own initiative, by responding to the latest achievements of the others, and the completion of their joint task will be greatly accelerated. We have here in a nutshell the way in which a series of independent initiatives, are organised to a joint achievement, by mutually adjusting themselves at every successive stage to the situation created by all the others who are acting likewise.[35]

Polanyi likens this self-coordination of independent initiatives to the workings of an invisible hand, where each individual scientist, by pursuing their own interests, contributes to the greater good of jointly discovering a hidden system of things. Polanyi is at pains to stress that self-coordination is far more effective than any form of central coordination:

> It is also obvious what would happen if someone tried to improve matters by applying central administration. Each helper would then have to await a decision taken at the supreme level. In effect, all of them except

34 See Polanyi, 1947, 1962.
35 See Polanyi, 1962, p. 55.

> the head of the organisation would cease to contribute to the piecing together of the puzzle. Cooperation would fall to zero.[36]

Knowing that Polanyi was a Jewish immigrant to the U.K., fleeing from Nazi persecution in 1933, and that he proclaimed the moral bankruptcy of Marxism, the villain in his story should be clear: the centrally planned science of totalitarian regimes (including the Nazi Regime and the Soviet Union), as well as its (Western) acolytes (including Polanyi's arch enemy, the British Marxist John Desmond Bernal).[37]

Central planning would be effective only if progress could be formalized. But it cannot, says Polanyi—not in the context of the jigsaw puzzle (for it is assumed that the end-result is unknown), nor in the context of science (where the end result is also unknown). Given this lack of foresight or "View from Nowhere", Polanyi tells us, we can only rely on the local, personal, and tacit judgments of individual puzzle-solvers and scientists.[38]

What do these judgments comprise in the context of science? How does mutual adjustment work there? Basically, individual scientists adjust to one another by taking note of the published results of fellow scientists. They generally contribute to the resolution of problems that they judge to be close to their expertise, excite them, and are neither too hard nor easy. But, for there to be progress, they cannot contribute just anything. Their work needs to be sufficiently plausible, have sufficient scientific value (viz., accuracy, systemic importance, and intrinsic interest), and sufficient originality—where the former two invite conformity, and the latter enforces dissent.

Like Popper and Kuhn, Polanyi thus believes that the growth of scientific knowledge results from a delicate balance between conservatism and novelty-seeking. What, then, would happen in a neomanic regime? (Note that this question doesn't arise for Polanyi, because he thinks that scientists tend to be conservative.)[39] There is a considerable risk that neomania would drive each individual scientist to pursue their own, self-defined puzzle—each puzzle disconnected from the other. Framed in those terms, the challenge posed by science's

36 See Polanyi, 1947, p. 583.
37 See Nye, 2011.
38 See Hartl, 2012, p. 314.
39 Ibid.

obsession with the new is to get scientists to agree to working on the same puzzle. And this calls for a different kind of coordination.[40] I will call it Peirce-style coordination, after the American scientist and father of pragmatism, Charles Sanders Peirce.

As Peter Galison and Lorraine Daston carefully detail, Peirce served as an aide at the U.S. Coast and Geodetic Society and actively contributed to the *Internationale Gradmessung,* a project begun in 1886 aimed at determining the exact shape of the Earth by measuring variations in the gravitational constant all over its surface.[41] Basically, the idea was to create a global network of local observers who would measure gravity at locations all over the world, and to unite these observations into a single Earth map. For the local observers to be coordinated into a global network, these observers would need to agree not just to set aside their own pet projects to partake in this project, but also on the units of measurement (absolute or relative gravity?), instruments (automated or observer-run?), required skills (high skill with higher accuracy but lower coverage, or low skill with lower accuracy but higher coverage?), etc. That is, the local observers would have to sacrifice many of their own epistemic idiosyncrasies—inhered in culture (e.g., national traditions of instrumentation) and the individual (e.g., differences in tacit knowledge, or in commitment to the project)—for the sake of the epistemic interests of the community. And, indeed, it took decades and dozens of conferences to wash out these differences; but, eventually, the project was a success,[42] until it was terminated by the start of World War I.[43]

Peirce extensively analysed the epistemic aspects of the *Gradmessung,*

40 Framed differently, by assuming that scientists are working on the same puzzle, Polanyi is assuming something that his story was supposed to show to be unnecessary.

41 See Galison & Daston, 2008.

42 At least epistemically speaking. Indeed, the stakes of the *Internationale Gradmessung* weren't merely epistemic; miners and colonizers, for instance, would—and in fact did—also benefit from the project. Peirce, however, seemed to have been primarily driven by epistemic interests. Although he acknowledges the utility of the project, he warns against "too great a stress [...] on the demands of the practically useful" (Peirce 1882, p. 507; quoted in Galison & Daston 2008, p. 309). In any case, my aim is not to portray the *Internationale Gradmessung* as a model for contemporary coordination. Rather, I want to use Peirce's reflections on the project to derive a couple of general principles that manifestly differ from Polanyi's principles, and that I think contemporary science would also profit from.

43 See Torge, 2012.

to the extent that Galison and Daston refer to him as a philosopher of scientific coordination.[44] First, one thing that stood out for Peirce was the primacy of the scientific community and what he called the 'self-denial' on the part of the individual scientist that this implicitly entails. So, while Polanyi holds that the common good is best served when individuals pursue their own interests, Peirce-style coordination relies on subordinating many of those interests to the greater good. Reality, according to Peirce, can only be revealed by transcending what he called the "vagaries of me and you"; only a coordinated community of observers is able to cancel out individual idiosyncrasies and the errors of single measurements.[45]

Second, Peirce saw self-denial operating also at the level of nations; they too needed to set aside their national pride and ambition for the greater good, and to practice what he called "international solidarity". In the context of the *Gradmessung* this meant, for example, that nations were to calibrate their instruments in one and one country alone (viz., Germany, particularly Berlin).[46]

Third, coordination was not imposed from above by any central government—the failure of the Napoleonic Empire to impose the metric system in its dominions led Peirce to believe that any attempt at doing so would have been futile.[47] Rather, the *Gradmessung* was coordinated bottom-up; it was scientists and not a central governor who, through numerous conferences and extensive correspondence, collectively reached an agreement on how best to organize the effort.[48] (The *Gradmessung* thus demonstrates that the dilemma Polanyi presents us with—namely, a dilemma between central coordination and self-coordination—is a false one.)

Finally, Peirce saw a new type of scientist at work in the *Gradmessung*—and he thought that this new type should become the norm.[49] It was a scientist who internalized an ethos that combined self-denial at

---

44 See Galison & Daston, 2008, p. 323.

45 Ibid., p. 324.

46 Ibid., pp. 325–326.

47 Ibid., pp. 326–327.

48 Bigg et al., 2023 argue that *Gradmessung* conferences were as political as they were scientific. For instance, participants represented their states, and governments substantively contributed to agenda setting and decision making. Be that as it may, there is no sense in which the effort was coordinated in a centralized way.

49 Ibid., pp. 327–328.

the service of the common good (as per the first point above) with a reluctance to yield to external authority (as per the third point above).[50] Such scientists had to be cultivated by incorporating into their training a *moral* training—or so Peirce believed.

To Peirce's observations let's add a couple of others (some of which may already be implicit in the above). The *Gradmessung* was special in its scope. The project engaged resources and perspectives from the entire community of geodesic metrologists. So in Peirce-style coordination, research priorities, methodologies, standards, and so forth are set through community-level deliberation; Peirce-style coordination thus goes beyond the coordination happening in local—or, in the case of the *Gradmessung*, national—teams.

Moreover, whereas Polanyi-style coordination leaves implicit the common epistemic good, Peirce-style coordination defines it in advance (and perhaps redefines it along the way), and collectively so—in virtue of negotiations at conferences and through other forms of exchange. So before embarking on the *Gradmessung*, the scientific community had to and did agree that the project would be not just worthwhile to pursue, but had to be prioritized over other projects.

Lastly, Peirce-style coordination is an *active* process. Coordination doesn't simply emerge, as Polanyi thought it did, through the workings of an invisible hand; it requires sustained interventions (e.g., negotiations pertaining to goal-setting and standardization) to bring and keep in alignment the activities of individual scientists, such that these activities effectively result in the attainment of the common goal. Indeed, Peirce actively participated in numerous conferences organized to initiate and sustain the *Gradmessung*.

Beware! Peirce-style coordination *can* but *doesn't need* to lead to big team science consortia

The *Gradmessung* was among the first, but certainly not the last, of scientific endeavours coordinated in a Peirce-style manner. In the 1950s, physicists came to realize that further advancements in their field would

---

50 But see Chapter 1 f.n. 43. So what Peirce recommended regarding not yielding to external authorities was perhaps more aspirational than real.

require budgets that single countries would not be able to cover, let alone single institutes.[51] So, in 1954, CERN—or the Conseil Européen pour la Recherche Nucléaire, or the European Organization for Nuclear Research—came into being.[52] The organization initially pooled resources from twelve Western European countries, but now comprises twenty-three member states. It operates the CERN laboratory (the largest of its kind) which includes, most notably, the Large Hadron Collider in which Higgs boson particles were discovered.

Biology followed suit in 1990, with the start of the Human Genome Project.[53] The project was a collaboration between twenty research institutes, located in seven countries, and by 2003 it managed to decipher the entire human genome sequence. Around the same time, behavioural geneticists started to pool their financial and intellectual resources, leading to large consortia such as COGENT (the Cognitive Genetics Consortium) and the SSGAC (Social Science Genetic Association Consortium). These collaborative efforts resulted in datasets comprising data on hundreds of thousands rather than dozens of individuals, and enabled relatively quick but careful study of, for instance, associations between genetic variants and educational attainment.[54]

Table 5.1 provides examples of more recently established big team science consortia. Many of these are grass-roots initiatives that, in effect, emerged in response to the concerns raised by the replication and generalizability crises.[55] They are all largely coordinated in a Peirce-style fashion. For one, they are all community-spanning, global initiatives (as indicated by the last column in Table 5.1). For another, they all require participants to subordinate their own interests to those of the collective—at least more than if they were to go it alone. By way of illustration, the Psychological Science Accelerator (PSA) is a globally distributed network of researchers (2468 researchers from 73 countries) that pools intellectual and material resources to accelerate the accumulation of rigorous knowledge in psychological science. In one of its studies, it tested whether Covid-19-related emotion-

51 See Forscher et al., 2023.
52 See Krige, 2004.
53 See Hood & Rowen, 2013.
54 See Trampush et al., 2017.
55 Taken from Coles et al., 2022a.

regulation strategies had any effect on subjects' attitudes and well-being.[56] Out of a concern for generalizability, it pursued this question globally: the PSA managed to mobilize no less than 450 researchers, who studied over 20,000 subjects from 87 countries. This study could not have been completed without substantive self-denial on the part of individual scientists (in Peirce's sense). For instance, researchers had to give up a possible first-authorship for an honourable mention in a list of 450 co-authors; and, before that, they had to be coordinated into foregoing their own idiosyncratic ways of working and instead aligning themselves with whatever procedures the community had jointly decided upon (regarding, e.g., study design, data collection, measurement tools, analysis, dissemination, and publication strategies).[57] This points to another relevant feature of the PSA: like the *Gradmessung*, its coordination is active and bottom-up—it has what it calls "decentralized authority".[58] So, as much as possible, members collectively decide on the direction of the PSA—i.e., on the policies it should implement and the projects it should prioritize. The ManyBabies consortium has a similar governance structure: all its project decisions (decisions regarding, e.g., research questions, study design, stimulus creation, piloting, data collection, analysis, interpretation of the findings, and writing) are made through consensus formation.[59]

Table 5.1. Examples of big team science research consortia. Coles et al. (2022a).

Consortium or Project Name	How and When Organized	Example of Project Finding or Question	Data Collection
Psychological Science Accelerator	Grass-roots consortium launched by a 2017 blogpost. Now involves some 1,200 researchers.	'Cognitive reappraisal' improves emotional reactions to the COVID-19 pandemic.	Data from more than 20,000 people in 87 countries collected by more than 450 researchers.

56 See Wang et al., 2021.

57 According to some insiders (personal communication with members of the PSA), consensus formation in the Covid study was less than perfect. The study would have been primarily designed by a few individuals, so that later participants had only two options: to join the study *as is*, or not to join. If so, it would not fully conform to the characteristics of Peirce-style coordination. This will be covered later in the main text.

58 See https://psysciacc.org/about.html

59 See https://manybabies.org/about/

Consortium or Project Name	How and When Organized	Example of Project Finding or Question	Data Collection
ManyBabies Consortium	Grass-roots consortium launched by a 2015 blogpost. Now involves around 450 researchers.	Infants prefer 'baby talk' even when it's not in their native language.	Data from 2,329 infants collected by 150 researchers in 16 countries across the world.
ManyPrimates project	Grass-roots consortium launched through 2018 symposium, word of mouth, e-mail and social media; now involves about 150 researchers.	Among 41 closely related primate species, phylogeny matters more for short-term memory than do ecology or social factors.	81 researchers studied 421 primates.
Nutrient Network (NutNet)	Launched in 2006 through e-mail and Twitter requests to join the network. Data collection began in 2007.	Does herbivory and light availability influence the loss of plant species caused by nutrient addition?	Using data from the broader Nutrient Network experiment (>130 collaborating sites), researchers documented effects of combined nutrient addition and herbivore exclusion on plant diversity at 40 sites across the globe.
Many Smiles Collaboration	Grass-roots effort launched in 2018. Collaborators formed smaller teams recruited through social media and e-mail.	Does changing facial expression affect emotions?	Nearly 50 researchers collected data from 3,878 participants across 19 countries.
Reproducibility Project: Cancer Biology	Launched in 2013 through funding provided by Arnold Ventures to the Center of Open Science and Science Exchange.	Can the results of experiments from high-impact cancer-biology papers be reproduced?	200 collaborators attempted to replicate 158 effects from 50 preclinical experiments.
Disturbance and Resources Across Global Grasslands network (DRAGNet)	Grass-roots consortium conceived in 2018. Data collection began in 2019; network built through e-mail and Twitter.	When grasslands are disturbed by tilling and nutrient additions, how do they respond?	Some 90 researchers monitor 70 sites in 18 countries (https://dragnetglobal.weebly.com).

Other consortia in Table 5.1 are more hierarchically organized. For example, the NutNet and DRAGNet consortia explicitly ask new members to conform to the research questions and research protocols that their first members have settled on; new members are thus excluded from community-level consensus formation. These, then, are imperfect instantiations of Peirce-style coordination (see also f.n. 55). Although they perform well when it comes to addressing reproducibility and generalizability, my insistence on consensual decision-making is motivated by other concerns: as will become apparent, community-level deliberation helps to safeguard (epistemic) diversity.

Importantly, Peirce-style coordination can, but does not necessarily, result in the formation of big team science consortia. Consider the following examples.

Every decade the U.S. Particle Physics Project Prioritization Panel (P5) meets to set the research agenda for particle physics for the decade to come, and advises science funders on how best to support the realization of that agenda.[60] The same happens in Europe: every so often, the European Strategy for Particle Physics (ESPP), through a broad consultation of the grass-roots particle physics community, solicits the opinions of physicists from across the globe to identify the priorities for the long-term future of the field. The process is closely coordinated with similar processes in the U.S. and Japan, in order to ensure optimal use of resources globally. Many of the process' recommendations are directed at science funders.

At conferences of the World Climate Research Program (WCRP), researchers (sometimes over 1,900) collectively reflect on future goals and developments to follow, resulting in an implementation plan that researchers worldwide try to align their research activities with.[61]

In rheumatology, the OMERACT initiative has, for over thirty years, deployed biannual consensus conferences to engage an international crowd of researchers and patients to prioritize research and improve methodologies.[62] So, for instance, regarding shoulder disorders, OMERACT 2018 participants decided with 71% agreement that the following domains would need to be covered in *any* clinical trial: pain, function, a global rating of the shoulder, and adverse events including

---

60 See Garisto, n.d..

61 See Asrar et al., 2013; Busalacchi & Asrar, 2009.

62 See Fried et al., 1993; L. J. Maxwell et al., 2019; Tugwell et al., 2007.

death; that four further domains could optionally be included: participation (in recreation or work), sleep, emotional well-being, and condition-specific pathophysiological manifestations; and that cognitive dysfunction didn't need to be included.[63] In the next phase, the OMERACT methodology was used to determine, for each domain, key measures (quantitative: e.g., measures of blood pressure, laboratory values, physical performance; qualitative: e.g., measures of quality of life, patient satisfaction) and measurement instruments—leading to standardization of measurements across trials and facilitating the accumulation of knowledge.

These four initiatives—P5, ESPP, WCRP, and OMERACT—embody key elements of Peirce-style coordination: a clearly defined common good, substantive subordination of personal preferences to that common good (including standardization), active rather than emergent coordination, engagement of the entire research community, and an international scope. However, unlike big team science consortia, they lack a formal management structure. Once consensus is reached, local teams independently determine whether and how to contribute to the agreed-upon goals, working autonomously without oversight from a governing body (see also the proposals in Chapter 6, in particular Figure 6.2).

Peirce-style coordination doesn't automatically lead to big team science in yet another sense. Given its scope, one might think that it is fit for highly complex and global puzzles but unfit for puzzles that are, or at least seem, local and not so wicked. If so, I would be mistaken in advertising Peirce-style coordination as a *general* model for how to re-organize large parts of science. Yet, Peirce-style coordination leaves room for local and explorative research, and for minority approaches. If the community deems it important to address, say, the high maternal death rate in India, wishes to explore a new avenue of research, or wants to assess the merits of a minority perspective, it can decide to invest some of its resources in those particular projects—for what Peirce-style coordination requires is merely that research priorities, methodologies, and the like are determined at the community-level.

63 See Ramiro et al., 2019.

Peirce-style coordination 2.0, and objections to the idea

Peircean self-denial is self-denial *vis-à-vis* the scientific community. Stengers, however, recommends self-denial *vis-à-vis* a broader community, one that includes relevant societal, non-industrial interest groups.[64] Recall why this is important. Engaging non-academic stakeholders other than industrial partners helps to relieve some of the pressure, exerted by industry, to rush research. And, because it—by design—aims to align research outcomes with broad societal interests, this contributes to addressing the applicability crisis.[65] Let's call Peirce-style coordination that manages to engage extra-academic, non-industrial interest groups 'Peirce-style coordination 2.0'.

Out of all the examples we have considered, only the OMERACT initiative is coordinated in Peirce-style 2.0. Ideally, all others would be, even when it comes to *basic* science projects. Organizing this may be complicated, but that is an issue I will take up in the next Chapter. First, I need to rebut four objections against rolling out Peirce-style organization—1.0 and 2.0—across science.

Objection 1: Academic freedom

The first objection concerns academic freedom. Polanyi's self-coordination results from leaving individual scientists to follow their own bent. For coordination to run smoothly and efficiently, Polanyi says, scientists ought to be given the freedom "to choose one's own problem of investigation, to conduct research free from any outside control, and to teach one's subject in the light of one's own opinions".[66] Peirce-style coordination, however, *does* considerably limit individual discretion, for it requires self-denial: to no small extent, scientists ought to align themselves with the wishes of many others—other scientists (Peirce-style 1.0) and the public (Peirce-style 2.0).

64 Stengers is not alone in this. See also, e.g., Douglas, 2005; Kitcher, 2011.

65 Such so-called *trans*-disciplinary research has many more advantages, advantages that are unrelated to solving the scientific crises, including the incorporation of local, non-academic knowledge, capacity-building among non-academic stakeholders, stakeholder ownership of the results of research, etc. See, e.g., Tebes & Thai, 2018.

66 See Polanyi, 1947, p. 583.

Polanyi's insistence on academic freedom is motivated by political concerns. As hinted above, he fears—and, given what he had seen happening in totalitarian regimes, rightfully so—that without such freedom, the state would decide what to investigate, how to investigate it, and what should be the result of the investigation.[67] With his defence of academic freedom, Polanyi places himself in a long tradition—starting at least as early as 1810, when Wilhelm von Humboldt made academic freedom the cornerstone of the modern university.[68] According to von Humboldt, the university should be a place where students and researchers can think freely, independently from authority, tradition, and dogma.

But there is no sense in which Peirce-style coordination 1.0 and 2.0 subject scientists to the authority of the state (or to tradition and dogma, for that matter). Rather, scientists put themselves at the service of the common good, which is determined not by politicians or ideologists but, jointly, with fellow scientists (1.0) and the wider public (2.0).[69] Further, participation in Peirce-style initiatives is voluntary; no one dictates what to think about, nor how to think.[70] Surely, when a scientist agrees to participate in a Peirce-style project, their hands may be tied more tightly than if they were to work Polanyi-style. Still, if political or religious interference is the principal worry, nothing militates against such a restriction of freedom.

Interestingly, Peirce too saw nothing inconsistent in advocating, at the same time, Peirce-style coordination *and* academic freedom. About the latter, he writes that "the health of the scientific communion requires the most absolute mental freedom".[71] So it is with full mental freedom that scientists must decide whether to subscribe to any consensus that may be formed—e.g., regarding the coordination of a project or regarding what to believe.

---

67 See Hartl, 2012.

68 See Fuller, 2009.

69 Accordingly, state-driven, mission-oriented research—such as the Apollo and Manhattan programmes—do not qualify as *Gradmessung*-style coordination.

70 This is not to say that there is no pressure whatsoever on scientists to participate, at least if the alternative is receiving no funding. That is why, in the next chapter, I write that scientists also need to be provided with a basic income.

71 See Peirce, 1931, p. 2220.

Objection 2: Diversity

One motivation for academic freedom is that it insulates science from political and religious interference; another is diversity—and this forms the second objection I need to rebut. The worry is twofold. First, although it still allows for relatively small local and explorative projects, Peirce-style coordination in actuality *does* reduce the number of research questions that science can meaningfully pursue. Second, Peirce-style coordination diminishes epistemic diversity, i.e., diversity regarding the possible perspectives and approaches one could take to any given question. More specifically, there is a risk that, within a project, during consensus formation, minority perspectives will be overruled by the tyranny of the majority. And there is a risk that the perspectives of those who chose or are forced to go it alone will get marginalized even more than they are today. Let's consider both worries in turn.

We would indeed need to accept that science can concern itself with a smaller number of research questions. But there is much to say in favour of robust knowledge about a little over unreliable knowledge about a lot, in particular when the robust knowledge pertains to issues that the community—of scientists (Peirce-style 1.0) or of scientists and the public (Peirce-style 2.0)—deems particularly important.

The problem of the tyranny of the majority is an issue that has been acknowledged by scientists participating in big team science consortia. Indeed, WEIRD people are over-represented in the governing and steering bodies of ManyBabies, the PSA, the Nutrient Network, and DRAGNet.[72] Very few of their publications have as the first author a researcher affiliated with an institution in a developing nation.[73] The reasons for the exclusion of researchers from under-privileged countries include a lack of infrastructure for them to participate in big team science projects, a lack of resources for volunteer work (e.g., serving as a coordinator of a big team science project) and so forth. Accordingly, there is a risk that big team science primarily reflects the interests of only a small fraction of the research community, viz., the WEIRD fraction.

The question we need to ask, however, is whether big team science—or, rather, Peirce-style coordination—creates new equalities, aggravates

72 See Coles et al., 2022a.
73 See Silan et al., 2021.

existing inequalities, or 'merely' perpetuates them. I think the latter is the case; because even in the current regime, what qualifies as the dominant paradigm is largely determined by the research carried out in (the most) resourceful labs. In fact, if organized well, Peirce-style coordination may contribute to *reducing* inequality. For one, the pooling of resources makes it easier to effectively reach out to under-represented groups. So, the PSA, ManyBabies, DRAGNet, and Many Primates all have—admittedly, imperfect—policies in place that aim for inclusion of the Global South.[74] Further, Peirce-style coordination lowers the threshold for participation in the scientific debate. Recall that in a Polanyi-style regime, it is the published literature that determines what is part of the debate. Those who lack the resources to publish material that conforms to the dominant standards thus tend to get excluded. In Peirce-style coordination, questions about coordination (research prioritization, setting of standards, preferred methodologies) are just as much part of the debate as the discussion of scientific findings; and a large part of this conversation takes place outside of scholarly journals. Thus, participation does not require a costly publication record. At least in principle, a qualification as a scientist and a small budget—at least relative to the budget necessary for scholarly publishing—for online and offline communication should suffice to have a seat at the table. Obviously, we must not be naïve. The obstacles for under-represented groups are manifold. For many in the Global South, attending conferences and participating in online discussions remain luxuries they cannot afford (time-wise, money-wise, safety-wise, politics-wise). Also, deliberation is never free of power. No form of coordination will fully erase imbalances relating to (publication-based) reputation nor, more broadly, imbalances relating to resources, geography, etc. But this is not a reason to avoid inclusivity—and, as I will continue to argue in the next Chapter, Peirce-style coordination may be our best shot at it.

Let's turn to the possible marginalization of excluded approaches. This is worrisome, indeed. For epistemic diversity—the co-existence of several approaches—has many possible benefits: (1) it ensures that scientific ideas are rigorously put to the test, in the sense of being critically assessed from more than one perspective; (2) it helps to manage the

74 See Coles et al., 2022a.

unpredictability of scientific development by hedging the bets; (3) it facilitates the fruitful combination of different scientific approaches and ideas; (4) it leads to healthy competition among such approaches and ideas; and (5) it helps to satisfy the different aims attributed to science (e.g., simplicity, accuracy, fruitfulness).[75]

A first thing to note is that, as I have emphasized before, the outcome of Peirce-style coordination may be a community decision to pursue more than one approach. So while it likely excludes more than Polanyi-style coordination, Peirce-style coordination doesn't need to result in the cultivation of one and only one approach.

Moreover, Peirce-style coordination allows for so-called adversarial collaborations (see, e.g., the Many Smiles Collaboration).[76] In such collaborations, rival approaches directly compete against one another, in an attempt to identify whose viewpoint is ultimately the most adequate regarding the issue under dispute. Typically, researchers with differing ideological or theoretical views team up, jointly determine in advance the best research protocol to settle their disagreement, and subsequently carry out the plan, often under the supervision of a neutral moderator. Adversarial collaborations thus *actively* contribute to bringing forth some of the benefits of epistemic diversity—most notably, benefits (1) and (4).

Finally, my call for Peirce-style coordination is a call for having *more* of it, not for completely organizing science according to its principles. This point is important also for addressing the next objection.

Objection 3: Conservatism

Research suggests that smaller teams disrupt science and technology with new ideas, whereas larger teams are more conservative and tend to develop existing ones.[77] This may be due to the fact that, to satisfy the desires of more people, larger teams will be more likely to cater to the 'median desire'.[78] So, if science was entirely coordinated Peirce-style, and if such coordination primarily resulted in big team science, science would become too conservative.

75 For an overview, see Rolin, 2019; Van Bouwel, 2015.
76 See Coles et al., 2022b.
77 See Wu et al., 2019.
78 See Forscher et al., 2023.

Indeed, I have repeatedly said that for there to be progress, science needs to strike a delicate balance between novelty-seeking and conservatism, and that neomania has tipped the balance towards the former. I then introduced Peirce-style coordination as a means for restoring the equilibrium. Obviously, we must avoid tipping the balance too far. So while we do need to attract more people to the RED AREA of Figure 1.2 in Chapter 1, we still need people to operate in the lower-left corner, producing novel ideas.

Let's first get one confusion out of our way. As I suggested earlier, Peirce-style coordination in actuality does leave room for small-scale initiatives. The community may for instance decide to spend a limited amount of resources on exploring a new area of research. Or, a small group of researchers may first identify problems and approaches unconstrained from a broader team consensus; subsequently, they may draft a proposal on the basis of their ideas and submit it for consideration by the larger community.[79]

But, arguably, this is not enough to safeguard novelty. The coordination costs of Peirce-style coordination may impede creativity. Further, novelty-seeking requires persistence. The literature on the delayed recognition of important scientific discoveries makes abundantly clear that individuals or small teams should not give up their pursuit of a new idea just because the (inherently conservative) community doesn't see its potential—it took researchers *decades* to accept, e.g., Mendelian genetics, continental drift, Pasteur's fermentation theory, the idea of *Australopithecus* being ancestral to *Homo*, and the viral origins of tumours. Community resistance to novelty is, of course, not only negative: among other things, it contributes to avoiding mega-mistakes—i.e., investing a lot of resources in the pursuit of novel but mistaken ideas.

All of this suggests that science should be partially coordinated, partially not: a significant portion of available funding should still go to independent science. But, because coordination is unlikely to monopolize science anytime soon—there is currently so little of it and the system is so inert—we can afford to primarily concern ourselves with getting more of it, as we will in the next Chapter.

79 Ibid.

Objection 4: Coordination costs

Coordination costs increase with team size. Consensus formation, in particular, is a resource-intensive process, especially when it is organized at such a high level as the community level. Polanyi-style coordination has no such 'overhead' costs. Its proponents may thus object to the inefficiencies of Peirce-style coordination.

But imagine what the *Gradmessung* would have resulted in if it had been coordinated Polanyi-style. Even if scientists had agreed on the goal—note that this, given Polanyi-style coordination, cannot be taken for granted—their idiosyncrasies would have prevented them from achieving it; they would have produced not a complete and coherent map of the Earth, but rather a haphazard collection of local maps. So, relative to the agreed upon goal, Polanyi-style coordination would be *less* efficient than Peirce-style coordination.

More generally, while Peirce-style coordination is costly, it is better than Polanyi-style coordination at reducing research waste—and it derives its efficiency from that. If each scientist follows their own neomanic agenda, we acquire bits of knowledge, each of which says little on its own. If the research protocol of the PSA's aforementioned Covid study had not been harmonized across all participating teams, the knowledge obtained would not have been robust because the results would be commensurably patchy. If biomedical researchers do not bring their research into alignment with patients' priorities, they risk producing irrelevant knowledge. And so forth.[80]

This is *not* to say that coordination costs can be safely ignored when promoting Peirce-style coordination. Far from it. We indeed must try to minimize them (and we may learn how to do this by looking at some success stories).[81] This *is* to say, however, that when assessing these costs, we need to assess them not in absolute terms, but rather against the background of the aims of science, viz., the production of robust knowledge.

80 In light of these considerations, I remain sceptical that arguments concerning the inefficiencies of big team science—when based solely on citation counts and impact factors, as reviewed by (Aagaard et al., 2020)—are suitable for capturing the dimensions that truly warrant evaluation.

81 See, e.g., Schimel & Keller, 2015, or the examples developed in Chapter 6.

Conclusion

Science should do less; but what it does, it should do better. It has to be more selective in the questions it addresses and spend more resources on the questions it selects. This will not happen if individual scientists continue to pursue their own, neomanic agendas.

At root, this is a coordination problem: the challenge is to get the interests of individual scientists to coincide with the interests of the community. Polanyi-style coordination will not give us that: the pursuit of individual interests leads to a proliferation of research *projects*, none of which receive the resources necessary to properly develop into a community-supported research *programme*—where a research programme comprises the collection of research activities represented in Figure 1.2 (i.e., the observation and identification of novel phenomena plus all RED AREA activities) in the pursuit of one or more community-determined research priorities. (Misunderstanding alert: here I thus substantively diverge from Imre Lakatos' original conception of the term *research programme*.)[82]

Peirce-style coordination fares much better in this regard: the community actively figures out what its principal interests are, and builds (not necessarily big) research programmes around those interests. Participants voluntarily align themselves with these programmes.

We need more of this. Making it happen is difficult, but not impossible—or so I will show in the next chapter.

82 See Lakatos, 1970.

6. Towards a science organized around research programmes

Scientific knowledge, we saw, grows through research programmes, not (just) isolated projects: important ideas should be given the opportunity to mature and develop, rather than be almost instantly given up just for the sake of novelty.

This chapter sketches a future in which science is in actuality organized around research programmes—surely not entirely, but at least more than it currently is. In this future, science funders incentivize researchers to engage in Peirce-style coordination, and to define and commit to a number of well-selected research programmes. Journals and universities do the same: they too change their incentive structure such that researchers are stimulated to take up research programme work—which includes all the epistemic activities in Figure 1.2 of Chapter 1.

Let's make all of this a bit more concrete. For each actor in turn, I first describe their current practices, then define what they could do differently, and finally consider and rebut a couple of objections.

Science funders

Current practice

Science funders currently launch calls for proposals according to disciplinary boundaries or broad thematic categories (e.g., Horizon Europe is built around several broad clusters including, among others, "Culture, Creativity and Inclusive Societies", "Digital, Industry and

 https://doi.org/10.11647/OBP.0507.06

Space", "Climate, Energy, and Mobility").[1] Researchers submit project proposals to their disciplinary panel, or to panels whose thematic focus aligns best with their own. Projects typically have a short duration of 3–5 years, and are carried out by teams that are relatively small (particularly relative to these teams' respective communities). Given these limited resources, these projects are sufficient to initiate but not to carry through a programme. Whether or not an initiated programme is continued after it ends is out of the hands of the science funder—and given neomania, it is likely that awardees will apply for subsequent applications with new ideas, and their peers will not pick up where the awardees left off. Programmes end prematurely.

Thematically, calls for proposals tend to be little constraining—there is plenty of room for applicants to exercise their academic freedom when it comes to topic selection. In a Polanyi-like fashion, science funders trust applicants and a handful of reviewers and panel members to be able to judge which sorts of ideas would benefit the community most, and thus which sorts of ideas deserve pursuit. Because there are usually no democratic checks and balances on the selection of panel members and reviewers, these judgments cannot be considered representative of how the community would prioritize research.

Standard funding schemes are highly competitive. In the case of the most prestigious ones—that is, personal grant schemes *à la* ERC Starting, Consolidator, and Advanced grants or NIH's R01 grants—there is little incentive for researchers to collaborate. In the case of team-based schemes, relatively small teams compete with other relatively small teams; science funders will not push them to further team up, not even if they propose a similar idea. From their own end, too, there is little reason to join forces with other teams. The recognition that researchers get is often proportional to the amount of external project money they bring in per capita; accordingly, even if success rates are low, it makes sense for teams to limit their size to the bare minimum.

Science funders predominantly work locally—they cater to national rather than transnational research communities even though many research questions, due to their nature or due to resource constraints, can only be effectively addressed by the latter. Indeed, some questions—

1 See https://research-and-innovation.ec.europa.eu/funding/funding-opportunities/funding-programmes-and-open-calls/horizon-europe_en

e.g., about human nature, global climate trends, the efficacy of health policies—require the engagement of research teams from all parts of the world; and these and many others questions require budgets that no single national science funder is willing to pay. With its Horizon Europe programme, which supports transnational collaboration, the European Commission may be an exception. Still, the Commission, too, promotes research projects rather than research programmes.

Future practice

In my envisioned future, science funders promote Peirce-style coordination by: (1) facilitating bottom-up research prioritization and research programme development through the organization of consensus conferences; (2) providing support for the realization of selected research programmes; (3) thinking and acting internationally.

(1) Science funders organize consensus conferences according to disciplinary or thematic categories, which aim to reach agreements regarding the research programmes that deserve to be prioritized and the strategies to realize selected programmes. To illustrate what a programme may involve, psychologists and relevant stakeholders might come to a consensus that there is an urgent need for effective psychological interventions combatting eco-anxiety. They may also agree on which interventions and (sub-)populations should be prioritized for (initial) study, the primary outcome variables, the standardization of measurements and analyses, the methodologies that still need further development, strategies for taking experimental results out of the lab, and so forth.

As noted in Chapter 5, consensus conferences have been effectively utilized in fields such as particle physics, climate change research, and rheumatology. In rheumatology, the OMERACT methodology follows a three-stage process (see Figure 6.1). In the first stage, researchers and relevant stakeholders generate a list of candidate domains and candidate measurement procedures, through literature reviews, focus groups, and qualitative work (e.g., interviews) ("Generating" in Fig. 6.1). In the second, consensus methods are used to come to an agreement on the prioritization of domains and measurement procedures ("Agreeing" in Fig. 6.1). The third stage is the vote on the proposed prioritization ("Voting" in Fig. 6.1).

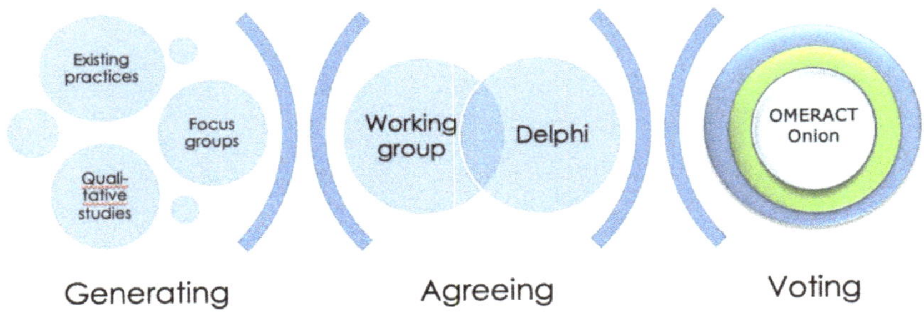

Fig. 6.1. The three stages of the OMERACT methodology. For a description, see main text. OMERACT (2021). OMERACT Handbook. https://omeract.org/wp-content/uploads/2021/12/OMERACT-Handbook-Chapter-Final-June_2_2021.pdf

(2) Science funders organize these conferences, with the promise that research proposals which align closely with selected research programmes have a high(er) chance of receiving funding. Funders may implement this in at least four ways.

In Option 1: Integrated prioritization, funders integrate research programme prioritization into their standard funding competitions. They issue calls for proposals based on disciplinary boundaries or broad thematic categories, as they currently do. However, proposals aligned with designated research programmes are prioritized, or at least given preferential consideration, over others.

In Option 2: Separate budget (all programmes), funders designate a separate budget specifically for research programme work, covering all selected programmes under a single allocation (so one funding scheme for all selected programmes). Only small independent team applications proposing to work on these selected programmes qualify for funding. This approach mirrors NWO's recently reinstated "Replication Studies Scheme" (see Chapter 2), which focuses solely on funding replication studies. However, it expands the scheme's scope to include RED AREA (see Figure 1.2) work beyond replication, and is targeted at designated programmes rather than being open-ended.

In Option 3: Dedicated budget per programme, funders allocate substantial, long-term budgets for each selected programme, with a dedicated budget per programme (so one funding scheme per programme). Independent research teams can continuously apply for funding throughout the programme's duration (e.g., ten years) to carry out specific parts of it. This sustained funding ensures that, eventually, significant parts of the programme are effectively covered.

In Option 4: Big team consortia, funders also commit significant, long-term funding to each selected programme, but allocate money to research consortia rather than, as in Option 3, to independent research teams. These consortia are then tasked with executing significant portions of the programme. Only this option, thus, would lead to the support of big team science.

The options differ across several dimensions, making some more suitable for certain circumstances than others. One distinction lies in the intended recipients. In the first three options, funding is directed toward relatively small, independently operating research teams. These options are ideal for *small* research programmes (which individual teams can fully undertake) or large *modular* programmes (where teams can handle one or more well-defined modules). Option 4, in contrast, is designed for big team consortia and is better suited for large and/or non-modular research programmes—those requiring active coordination of interdependent elements—or programmes involving essential yet routine tasks that otherwise might not be taken up. However, Option 4 comes with higher coordination costs and reduced flexibility, as programmes cannot be easily discontinued. If these constraints are a concern, funders deeply committed to advancing research programme work and addressing science's poly-crisis might favour Option 3, where teams operate without centralized oversight, and budgets can be revoked at any time—e.g., when a programme turns out to be misguided. Meanwhile, Option 1 and Option 2 are well-suited for funders looking to experiment with increasing their support for research programmes in a controlled, low-risk manner.

This list of options is not meant to be exhaustive—I am sure others exist. However, my aim has been to highlight the potential directions funders can take to make research more programme-centred and to illustrate some of the key considerations in selecting the option(s) that best align with their priorities.

(3) One of the success factors of the *Gradmessung*—as well as of OMERACT and the big team science consortia of *Chapter 5*—is their international outlook. As noted earlier, many scientific questions transcend national boundaries. In light of this, the common practice of using these boundaries as implicit funding criteria is highly artificial. When researchers are willing to collaborate across countries on such 'universal' questions, national funding agencies should make a serious effort to support and facilitate such collaboration. This likely means that science funders, not just scientists, would need to cooperate on a much larger scale than they currently do.

CONSENSUS CONFERENCE

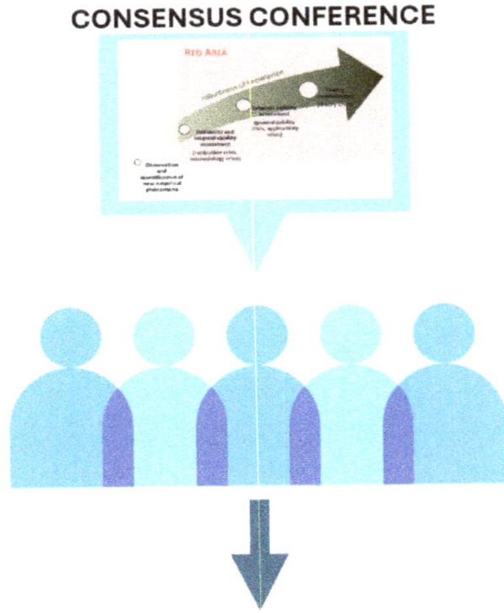

Option	Description	Key Differences	Benefits	Downsides
Option 1 – Integrated Prioritization	Aligns with existing calls; proposals matching selected programs get preferential treatment.	No separate budget; uses standard competitions.	Low risk, easy to implement, flexible.	Limited impact; weak incentive for program alignment.
Option 2 – Separate Budget (All Programs)	One dedicated scheme for all selected programs combined; independent teams apply.	Separate funding stream, but shared across programs.	Stronger incentive than Option 1; still relatively low risk.	Competition between programs; limited funds per program.
Option 3 – Dedicated Budget per Program	Long-term funding for each program; independent teams apply continuously.	One scheme per program; sustained funding.	High flexibility; modular approach; can discontinue programs if needed.	Requires significant resources; more administrative complexity.
Option 4 – Big Team Consortia	Long-term funding for consortia to execute large parts of a program.	Centralized execution; high coordination.	Enables large-scale, interdependent projects; strong commitment.	High coordination costs; low flexibility; hard to stop if misguided.

FUNDING OPTIONS

Fig. 6.2. Schematic representation of the proposal for funders. Funders first organize consensus conferences, where researchers and other stakeholders, in a bottom-up fashion, prioritize and develop research programmes (see Fig. 1.2). Funders subsequently support work on selected programmes in either of four ways (Options 1–4). Image by the author.

Once again, funders would have several options. National agencies could establish research-programme-specific alliances with other national agencies to jointly organize consensus conferences. For highly non-modular research programmes, they might consider pooling resources to fund transnational big team science initiatives. In contrast, for more modular programmes (such as the *Gradmessung*), agencies could maintain separate budgets, each funding only the work conducted within their own national territory. Additionally, a portion of the total budget could be pooled and allocated to support research teams with access to critical populations but lacking sufficient funding—such as teams in underprivileged countries. And so on.

Objections to the proposals

In the previous chapter, I identified and rebutted four *general* objections to Peirce-style coordination (viz., academic freedom, diversity, conservatism, and coordination costs). Here I identify and rebut several objections to the *specific* proposals I made above.

The proposals are too costly

One might worry that my proposals come with excessive overhead costs—expenses related to organizing, managing, and administering consensus conferences, the time researchers spend attending these events, the effort required to build transnational alliances, and so on.

This acknowledged, in response: the proposals also may *save* quite a bit in terms of resources. Consensus conferences can help to reduce the workload of grant reviewers, who will not need to invest as much time in assessing a proposal's general objectives, methodological choices, originality, and relevance, because these have been established in advance (i.e., at the programme-level). In the model where funders support big team science consortia (Option 4 above), only a handful of proposals would actually need to go through review—significantly easing the review burden of science funders, panel members, and reviewers. To put that burden into perspective, the NSF annually reviews about 50,000 proposals, and most of these are reviewed by multiple external reviewers. In 2005, the total time that all reviewers together spent on evaluating applications for the U.K. Research Council amounted to 192 years.[2]

What applies to reviewers also applies to applicants: they too can devote significantly less time and effort to defining the objectives, methodology, originality, and relevance of their proposals—often the most time-consuming and tedious aspects of grant writing.

Given the extremely low success rates of standard funding schemes—often 10% or lower—researchers waste vast resources on unsuccessful proposal writing. To illustrate, a study on the Australian National Health and Medical Research Council found that even with a relatively high success rate of 20%, unsuccessful applicants in 2012 alone lost the equivalent of over four *centuries* of research time.[3] As mentioned earlier, the core idea behind my proposal is to provide high(er) success rates to applicants who align with selected research programmes. Option 3 and Option 4 in particular would enable funders to considerably reduce the research waste associated with rejected applications.

Finally, even if my proposals were more costly than standard practices, they would likely make research allocation more cost-effective. Rather than producing isolated fragments of knowledge with uncertain reliability and significance, investments are intended to yield cohesive and dependable bodies of relevant research—unless, of course, programmes unexpectedly turn out to be 'mistaken'...

The proposals could lead to 'mega-mistakes'[4]

Scaling up to the level of supporting programmes increases the cost of errors. This is particularly concerning when funders choose to provide long-term support to big team science consortia (Option 4 above). In other options, budgets can be revoked whenever issues arise. However, in Option 4, challenges such as poor consortium management, unexpected scientific outcomes, interdependencies causing delays, and error propagation can significantly impact the programme. While

2 See http://www.rcuk.ac.uk/RCUK-prod/assets/documents/documents/rcukprreport.pdf

3 See Herbert et al., 2013.

4 A mega-mistake occurs when a research community spends a large amount of resources on a programme that later turns out to be mistaken.

similar issues can occur in smaller research projects, there they result in far less resource waste.

In reply, first, a major rationale for *public* funding of research is that governments can take risks that, say, for-profit organizations are not willing to take.[5] Accordingly, if governments take their role seriously, they must invest in research that is 'high risk/high gain'; they have a strong mandate to support research programmes, given such programmes' possible gains and despite such programmes' risks.

Second, as already noted, research programmes tend to be more conservative, and thus safer, than smaller projects. For one, they involve larger amounts of people—such that they are more likely to gravitate around a 'median desire'.[6] And they tend to build on existing ideas, rather than to generate from scratch (so to say) novel ones.[7] By implication, for research programmes to emerge, there is still a need for a flow of new ideas—and part of science funders' budgets would still need to go to that, preferably in the form of providing researchers with a basic income.[8]

Third, national funding agencies can distribute the risks of mega-mistakes by teaming up with other national funding agencies—so, by thinking and acting transnationally.

Fourth, consortia may implement additional internal and external quality checks. For instance, in CERN, any scientific manuscript is subjected to an iterative process of feedback, where all feedback must be addressed before submission as a pre-print.[9] Consortia may also deploy 'pilot testing', and formal methods that help to decide whether to stop data collection so as not to overinvest in bad ideas.[10] They may let their research process be monitored and assessed by external auditors, and perhaps even engage so-called 'red teams'.[11] Such teams serve as a designated 'devil's advocate', and are charged with finding holes and errors in ongoing work—much like 'white-hat hackers' in the software industry are charged with identifying security issues before they are

---

5 See Arrow, 1959; Veugelers et al., 2022b.
6 See Forscher et al., 2023.
7 See Wu et al., 2019.
8 See Vaesen & Katzav, 2017.
9 See Leogrande & Nicassio, 2021.
10 See Lakens, 2014; Schönbrodt et al., 2017; Forscher et al., 2023.
11 See Lakens, 2020.

discovered and exploited by malefactors. Red teams check every part of the research process, and consortium members need to incorporate their feedback. Red team members would need to be rewarded for their work—but, given the apparent conflict of interest, giving them co-authorship is not an option. So rewards need to be pecuniary, or based on increases in other types of recognition. We will return to the latter in a later section.

The proposals involve too much state interference

I previously argued that Peirce-style coordination aligns with academic freedom, because researchers voluntarily adhere to community consensus rather than state directives. But now that I have outlined how such consensus is formed, concerns about state interference may still arise. Given that coordination is essential for addressing science's poly-crisis, I suggested that governmental science funders take on the role of organizing and managing consensus conferences—without such involvement, the conferences might not happen at all. Suppose, then, that a research community reaches consensus on the need for a research programme on LGBTQ+ rights. The ultimate decision to fund such a programme rests not with the community but with the science funder, who—potentially influenced by political considerations—may ultimately choose to reject it. How to respond to this worry?

The most effective response begins by examining the typical structure and organization of governmental science funders. Agencies such as the U.S. National Science Foundation (NSF), the European Research Council (ERC), and the Dutch Organisation for Scientific Research (NWO) have a high degree of independence from government control. Rather than being overseen by a ministry or parliament, these agencies function largely autonomously. While their budgets come from government allocations, they retain full authority over how funds are distributed. Typically, funding decisions follow a multi-stage process. Independent scientists and experts conduct initial evaluations, followed by final approvals from the agency's higher management—officials who do *not* serve as government representatives. This independence is precisely designed to ensure that funding decisions are based on merit and peer review rather than political considerations.

To be sure, the independence is not absolute. As said, governments provide funding, and often have a role in appointing senior council officials. Still, they do not engage in the daily operations of these agencies nor interfere with individual funding decisions. Other agencies, such as the U.K. research councils and the U.S. National Institutes of Health (NIH), are accountable to specific ministries yet also maintain operational independence when it comes to funding allocations.

In sum, science funders have structural safeguards to minimize government interference, regardless of whether funding decisions concern individual projects (as is common practice) or larger programmes (as I propose). Given their scale, programmes may draw more attention from politicians, increasing the risk of intervention. However, potential misuse should not be a reason to reject my proposals. Rather than discarding them entirely, throwing out the baby with the bathwater, we should ensure—perhaps more than ever (see the book's Epilogue)—that these safeguards function as intended.

Scientific Journals

Current practice

Researchers submit manuscripts reporting on their study. In mainstream journals, these manuscripts are assessed by reviewers based on criteria such as novelty, estimated impact, methodological soundness, embeddedness in the extant literature, clarity, openness, ethical standards, etc. The criteria 'novelty' and 'estimated impact' militate against the acceptance for publication of studies that emerge from an ongoing research programme with prior publications. In any case, the two criteria are always assessed anew, that is, separately for each individual paper coming out of a research programme.

The publication process itself is also centred around individual studies. Journals collect accepted papers and compile them into issues, typically on a 'first come, first served' basis rather than grouping them thematically—except in the case of special issues. Even when multiple articles stem from the same research programme, it is left to the reader to recognize this.

More recent journals—such as the journals *PLoS ONE*, *Royal*

Society Open Science, and *Scientific Reports*—exclude 'novelty' and 'estimated impact' from their peer review criteria. Further, they do not organize articles into issues but instead publish them immediately upon acceptance. Still, these journals also do not explicitly link related studies—there are no designated 'streams' nor other mechanisms for connecting papers within the same research programme. The same applies to pre-print servers like *arXiv*, which categorize submissions by discipline but otherwise function as unstructured repositories where authors can drop and upload their papers.

As mentioned in Chapter 5, in *Scientific Utopia I*, Nosek and Bar-Anan propose several other modifications to current publication practices.[12] But these also maintain the individual study as the primary unit of peer review. Indeed, Nosek and Bar-Anan's suggestions—such as publishing peer review reports alongside articles and enabling post-publication review—apply to individual papers rather than research programmes as a whole.

Future practice

To incentivize researchers to take up RED AREA work, journals could do any of the following. Journals could: (1) accept, by default, all methodologically sound studies that build on previously published work and structure publications into dedicated streams, with each stream corresponding to a specific research programme; (2) allow the submission of programme-level (not just article-level) registered reports; (3) acknowledge research programme-level (not just article-level) contributions.

(1) If a journal considers a study important and decides to publish it, subsequent studies building on it—such as replications, generalizability checks, applications, or methodological developments—should no longer need to meet the same importance threshold. Instead, importance is assessed only once, for the original paper, and then automatically extended to later publications. This transfer makes perfect sense: an idea's significance grows in proportion to the extent to which it has been rigorously tested.

12 See Nosek & Bar-Anan, 2012.

In other words, as long as their follow-up studies are methodologically sound, authors are assured that their work will be published in the same venue as the original study—associated with their publication would thus be the *same* journal-based metrics (e.g., impact factor, Eigenfactor, journal ranking, prestige).[13] If journals still publish in paper issues and page limitations are a concern, subsequent studies could be published online only.

Online publication of articles would happen in streams anyway. Similar to how platforms like *arXiv* and *F1000* track different versions of an article, journals would maintain a record of all follow-up studies linked to each original publication. These articles would explicitly cross-reference one another, so that readers can evaluate the full body of evidence pertaining to a research programme.

(2) 'Registered report' is a publication format in which authors submit their study plan for peer review before conducting the research. If the plan is approved and followed through, the results are guaranteed publication, regardless of the outcome.

Initially, the format was designed for individual studies. But the format could be scaled up to cover research programmes. In this model, journals commit to publishing any results generated by programmes that have passed peer review—again, regardless of outcome. The peer review process for research programmes could take different forms, depending on how funders support programme-based work. For instance, if funders back big team science consortia (Option 4 above), these consortia could submit the same proposal they provided to the funding agency—journals might even base their decision on the referee reports already collected by the agency. In Options 1–3, programme descriptions could instead be informed by reports summarizing the outcomes of the consensus conferences where the programme was originally defined.

(3) Journals are increasingly adopting the CRediT system to assign authorship.[14] Only researchers who have taken up minimally one of the fourteen roles that CRediT distinguishes—"conceptualization", "data

13 Note that, for reasons given below, I think journal metrics should be abandoned as indicators of the quality of individual articles. So, this statement should be read as a concession to those who favour journal-based assessments.

14 See https://credit.niso.org/

curation", "formal analysis", "funding acquisition", "investigation", "methodology", "project administration", "software", "resources", "supervision", "validation", "visualization", "writing: original draft", "writing: review and editing"—are allowed to appear as (co-)authors. These roles primarily capture contributions to individual studies/ research articles. To incentivize scientific coordination and research programme work, journals may adopt a more inclusive approach to authorship. Researchers who played an active role in a consensus conference where a programme was designed meaningfully contributed to its 'conceptualization' and should receive formal recognition for that contribution—for instance, in the form of an authorship credit on all programme outputs. To promote actual RED AREA work, the same principle could apply to those doing such work. If this level of inclusion is deemed too broad, 'distant' contributors could be formally *acknowledged* in all programme outputs. Journals could enhance the visibility and standardization of acknowledgment sections (much as in the CRediT system) such that 'distant' contributions can be automatically indexed by platforms like ORCID. In the case of big team science consortia (see, again, Option 4 above), journals could, as some already do, accept consortium authorship—with all consortium authors appearing, in the appropriate order, just at the end of the article.

Objections to the proposals

Journals are too risk-averse to implement the proposals

Concerns about mega-mistakes are as pertinent to journals as they are to science funders—few journals will want to risk being associated with a failed research programme. And while public funders have a duty to take risks for the common good (see above), commercial publishers have no such duty. Their current business model, which revolves around the publication of isolated studies, works well. Accordingly, they have a vested interest in preserving the *status quo*.

Science funders may challenge this *status quo* by establishing their own publication process—one that integrates points (1) to (3) above. These efforts may be modelled on *Open Research Europe* and *Wellcome Open Research*. *Open Research Europe*, developed by the European

Commission, provides an open-access venue for research funded under the Horizon Europe framework. Manuscripts are pre-printed on the platform, undergo open peer review post-publication (focusing solely on methodology), and are published on a continuous basis. *Wellcome Open Research* is a similar platform, but caters to researchers who receive funding from the Wellcome Trust. These platforms are Diamond Open Access, meaning that they do not charge fees to either authors or readers—the publication costs are fully covered by the science funder.

Although these platforms already do much of what I think needs to be done—e.g., they are Open Access and assess contributions on methodology regardless of outcome—they could be more programme-oriented than they currently are. Implementing this shift would be relatively straightforward. Regarding (1), the platforms do already assign articles to thematic "collections" (*Open Research Europe*) or "gateways" (*Wellcome Open Research*). But instead, they could structure publication streams around research programmes and actively cross-reference related studies within the same programme. For (2), these science funder-managed platforms could accept programme-level registered reports, using as a basis the final reports from the consensus conferences they themselves have organized. And, when it comes to point (3), the platforms could take the lead in developing a system which ensures that researchers who engage in scientific coordination and programme-specific work receive proper recognition.

To further push commercial publishers to implement (1) to (3), science funders could *require* the researchers they support to publish their results in science-funder-managed Diamond Open Access outlets—and have this requirement in place at least until commercial providers have adapted their publishing models. Governmental grant-giving bodies wield significant influence—let them use it.

At least initially, researchers would no longer get credit for publishing in prestigious, high-impact commercial journals. But, to the extent that journal-level metrics are a good indicator of research quality at all (more on this below), publishing in, say, *Open Research Europe* or *Wellcome Open Research* would serve as a sufficient mark of quality. That is, researchers would gain recognition simply by securing funding for their work. The fact that a study is published in those publication platforms indicates that a large number of people have vetted its importance (*ex ante*) and that, additionally, it has passed a thorough methodology check (*ex ante* and *ex post*). Evaluation criteria for researchers—such as those used in hiring and promotion—would need to be adjusted too, but that is a matter we deal with later.

General worries about registered reports

Registered reports for individual studies have been criticized as overly rigid, demanding for reviewers, and time-intensive. These concerns apply even more strongly to registered reports for research programmes, particularly when these are undertaken by big team science consortia—as discussed in Option 4 above. Let's focus on that option.

When it comes to rigidity, research programmes are even harder to plan in advance than individual studies. Researchers may feel restricted by their initial proposals, with limited flexibility to explore unexpected findings, adjust methods as needed, adapt to changes in consortium composition, or respond to timing issues and other interdependencies. There are several—albeit imperfect—strategies to manage these risks. One approach is to keep the requirements for research programme descriptions minimal, outlining a general direction rather than prescribing a rigid step-by-step plan. Another is to structure programmes in a modular fashion, minimizing interdependencies to enhance adaptability. To account for the unpredictability of scientific research, certain modules can be designated as 'optional', 'exploratory', or 'contingent'. Additionally, programmes can incorporate detailed contingency plans, identifying potential risks, mitigation strategies, and alternative courses of action should those risks materialize.

I have previously addressed concerns about the demanding and time-intensive nature of peer review. While evaluating research programmes will undoubtedly require significant resources, it is a one-time process. As noted, the assessment conducted for funding decisions can also serve as the basis for publication decisions. Moreover, my proposals help conserve resources both upstream and downstream. Upstream, reviewers would evaluate only a limited number of programmes, while downstream, individual studies within a programme would require only a methodological check.

Universities

Current practice

Hiring and promotion

Hiring and promotion decisions are typically based on individual researchers' direct contributions to specific studies and publications. Indeed, widely used evaluation metrics—including the h-index, publication count, citation count, and Altmetrics (which track publication usage and discussions across social media and traditional media)—are all calculated at the individual level rather than at the programme or consortium level. Accordingly, the performance of other research teams within a programme or of the consortia to which researchers belong is overlooked.

Standard practice is individual-oriented in another sense. Few, if any, commonly used hiring and promotion criteria account for a researcher's commitment to scientific coordination and research programme work, including, but not limited to, consensus-building, research programme development, and consortium building and maintenance.

Another common practice is assessing researchers according to the impact factors of the journals they publish in. But publication venues that we earlier said to be most in keeping with the re-organization of science we advocate—viz., Diamond Access journals curated by science funders—do not have an impact factor, let alone a high impact factor. This standard practice, therefore, discourages researchers from publishing in these outlets.

Finally, academic hiring procedures typically rely on a CV, a teaching and research statement, an interview, a mock lecture, and a research presentation. They rarely include explicit assessments of soft skills, interpersonal and inter-cultural competencies, nor preferences related to work environment, tasks, and work style.

Education

I have used parts of this book in my teaching, particularly in a first-year MSc course on meta-science. Students are struck by the scale of the crises. They also appreciate the clear need for greater coordination in research. When reflecting on their education, they quickly recognize how much it emphasizes the ideal of the innovative individual. They note that, if disciplinary history is taught at all, it tends to centre on the groundbreaking contributions of a few prominent figures, that their BSc, MSc, and PhD theses must be completed independently, and evaluations of these works often prioritize novelty above all else. As they observe, independence of thought is valued more than a collaborative mindset, with little emphasis on the broader contributions they might make to the greater good.

All of my students regularly work on small-team assignments (typically 3 to 5 members), which is common at a technical university like mine. On multiple occasions, they have expressed feeling unprepared for this type of collaboration and have voiced a strong interest in receiving training that would enhance their teamwork, including training pertaining to project management, communication, negotiation, and conflict management skills. Unfortunately, I always have to tell them that there is no space for such training—neither in my course nor in their curriculum.

Future practice

Hiring and promotion

Universities can pull several levers to encourage researchers to engage in scientific coordination and programme/consortium work. Regarding hiring and promotion, universities can evaluate and reward candidates according to (1) programme-/consortium-level rather than individual-level performance criteria; (2) their contribution to programme/consortium development and management; (3) publication criteria that are programme/consortium-friendly. Further, when hiring, (4) universities can select candidates based on personalities, skills, values, and preferences that accord well with scientific coordination and programme/consortium work.

(1) Basically, programme-/consortium-level performance can be measured in any of the ways used to measure individual-level performance. Metrics such as the number of publications, citations,

Altmetrics, and the h-index are just as relevant for groups as they are for individuals. While these metrics are not yet widely available at the programme or consortium level, greater adoption of collective authorship could prompt science indexing platforms (e.g., Google Scholar, Scopus, Web of Science, ORCID iD) to adjust their systems accordingly. The same applies to qualitative criteria such as awards, policy impact, and community outreach. In principle, programme groups and consortia can be evaluated based on these measures, though some scientific traditions may need to evolve first. Prestigious awards like the Nobel Prize and the Fields Medal, along with national prizes such as the Spinoza Prize (Netherlands) and the Humboldt Award (Germany), still emphasize the notion of the solitary genius, perpetuating a distorted view of scientific achievement. These prizes are overdue for an update. If the Nobel *Peace* Prize can be awarded to organizations, why shouldn't major research prizes recognize the achievements of groups?

(2) The traditional academic CV is not well-suited for highlighting one's commitment to scientific coordination, whether in programme work, consortium efforts, or red team initiatives. However, assessment committees may request narrative CVs, which offer a more flexible format for showcasing these contributions. Unlike standard CVs, narrative CVs present a person's achievements and experiences in a story-like manner, allowing researchers to provide detailed evidence of their 'soft' skills. This format enables applicants to articulate their roles in coordination and teamwork—qualities that are often difficult to capture in a conventional CV.

(3) In 2012, at a meeting of the American Society for Cell Biology in San Francisco, the DORA (Declaration on Research Assessment) was issued. One of its most significant actions was denouncing the use of journal impact factor as a criterion for research assessment. DORA highlighted three key issues with this practice: (i) citation distributions within journals are highly skewed, meaning the citation rate of an individual paper has little correlation with the journal's impact factor; (ii) journal impact factors can be easily manipulated through editorial policies—journals can artificially boost their ratings by prioritizing review articles over primary research, exploiting publication bias, or forming citation cartels; (iii) the data used to calculate impact factors lack transparency because they are proprietary, controlled by commercial

organizations such as Thomson Reuters and Elsevier.[15]

Despite the fact that by now about 2,500 organizations—including universities, science funders, publishers—have signed the declaration, the journal impact factor remains a key metric for assessing research quality and impact—publishing in journals without an impact factor, such as *Open Research Europe* and *Wellcome Open Research*, is tolerated but not encouraged. That these two platforms are still unpopular as publication venues is suggested by their output so far. Since it was launched in March 2021, *Open Research Europe* has published just 779 articles—an average of 216 per year, despite Horizon Europe funding 5,000 projects annually.[16] *Wellcome Open Research* has published 1,743 papers (217 per year), but it started in November 2016.[17] By way of comparison, *PLoS ONE* publishes over 15,000 articles per year.

So unless science funders require the researchers they support to publish their studies on these funders' publication platforms (which may be a good idea anyway; see above), assessment committees should reward such behaviour—or any other type of publication behaviour that fits well with coordinated science.

(4) As we saw Peirce noted, scientific coordination demands a new kind of scientist—one who balances selflessness in service of the common good with a healthy scepticism towards external authority. But what else does it take? Most likely, it calls for teams made up of diverse individuals who bring different strengths to the table, rather than teams composed of individuals with a single ideal personality, skill set, or value system. Nonetheless, certain salient personalities, skills, and values may be under-represented in the current scientific workforce. For example, we may need more people who are eager and intrinsically motivated to conscientiously perform essential but routine tasks like replications, who prioritize consensus over conflict, who thrive in intercultural settings, who find fulfilment in group rather than individual achievement, and who have the patience that is required for slower forms of science. Typical assessment committees neither explicitly assess candidates based on these qualities nor actively seek individuals who possess them. Changing this would not only increase the likelihood

15 See https://sfdora.org/read/
16 Count made 6 December 2024.
17 Count made 6 December 2024.

that new hires will contribute effectively to programme work but also enhance their sense of fulfilment. A stronger sense of fulfilment, in turn, can lead to higher-quality work.

Education

If collaboration is essential to scientific progress, it must be central to students' curricula—starting with how we teach disciplinary histories. Newton famously acknowledged that he was not a solitary genius, stating, "If I have seen further [than others], it is by standing on the shoulders of giants".[18] Yet, students are rarely taught who these giants were, and when they are, the focus tends to remain on familiar figures like Copernicus, Galileo, and Kepler. But, as neatly documented by James Poskett, just as important for Newton was the indigenous knowledge of Incas, Polynesian tribes, and peoples living in the Arctics—knowledge brought to Europe by European explorers such as Charles-Marie de La Condamine and James Cook.[19] Disciplinary histories should be taught in a way that acknowledges the many (many) contributors who have shaped scientific fields, rather than reinforcing the myth of the lone genius.

In my first-year MSc course on meta-science, I challenge students to develop solutions for science's poly-crisis (without giving them access to this chapter, for obvious reasons). One of their favourite proposals is to make a meta-science course mandatory for *all* students—currently, it is required only for a relatively small educational programme. (And, for what it's worth, I don't get the sense that this suggestion is merely an attempt to appease me as their teacher.) I fully agree. As mentioned, the course effectively raises awareness of what science should ideally achieve—namely, robust knowledge—and highlights the need for RED AREA work, as well as greater collaboration and coordination, to reach that goal.

When it comes to BSc, MSc, and PhD research, a clear way forward is to integrate students into research programmes or consortium work, rather than having them conduct their own small, isolated projects. For PhD students, this could mean requiring that they devote at least part of their dissertation to such collaborative work. Assessment criteria should

---

18 See Newton, 1675, https://discover.hsp.org/Record/dc-9792/Description#tabnav

19 See Poskett, 2022.

also evolve, shifting the focus from novelty to a stronger emphasis on cooperative ethos. In this regard, students—like scientific staff—would benefit from training that prepares them for intensive collaboration.

Objections to the proposals

The proposals lack a robust empirical foundation

Throughout the discussion above, I have repeatedly attempted—but failed—to find solid empirical evidence in support of my proposals. While it seems intuitively plausible that, say, narrative CVs can effectively convey experience in scientific coordination, that education in collaboration and communication can effectively enhance students' and researchers' performance, and that hiring assessments can effectively help in identifying candidates well-suited for research programme work, the Science of Team Science (SciTS) is too young a field to have firmly established these claims. Although SciTS is evolving rapidly, many key factors—ranging from science policy to psychological and group dynamics—remain poorly understood. Evidence-based strategies for optimizing team composition, formation, and functioning are still in development.[20]

This should not be a reason for inaction, though. SciTS is still an emerging field, and like any evolving discipline, it requires ongoing experimentation and refinement. If we were to demand rigorous empirical proof before implementing any changes, we must also ask whether the current system has itself ever been rigorously validated. More often than not, existing academic structures and incentive systems are the product of historical convention rather than evidence-based decisions.

Further, research from related fields—such as organizational science, psychology, and management—already offers indirect yet meaningful evidence supporting the importance of collaboration and coordination, as well as strategies to enhance them. While more research is needed to tailor these principles to the academic environment, waiting for definitive proof before making adjustments risks paralysis.

20 For a review of the evidence, see Hall et al., 2018.

In light of this, I advocate for a measured, gradual approach to systemic change. Rather than implementing my proposed reforms—both here and in previous sections—all at once, we can introduce pilot initiatives and evaluate their impact systematically. The advancement of SciTS depends on researchers and institutions being open to experimentation, testing new approaches, and refining them as evidence accumulates.

Conclusion

I have sketched a future where science revolves more than it currently does around research programmes rather than isolated projects, in order to ensure the accumulation of robust knowledge. Achieving this shift requires changes across multiple levels: science funders must rethink how they allocate resources, journals must adapt their publication and recognition systems, universities must revise hiring and evaluation criteria, and academic training must better prepare researchers for collaborative work. For each of these actors, I have outlined various specific ways they could contribute to my envisioned future.

The sheer number of my proposals might give the impression that I advocate for rapid, radical change. But I do not. Science funders, journals, universities, and researchers must experiment with different models—hopefully drawing inspiration from my suggestions—evaluate their effectiveness, and fine-tune them based on emerging evidence. And they must remain alert to risks associated with bureaucratic overheads, mega-mistakes, and state interference.

This means that resolving science's poly-crisis will take time, no matter how much we wish otherwise. Still, given the progress the open science movement has achieved in just over a decade, a resolution need not take forever—provided we also commit to concerted action.

7. Epilogue: AI, the crisis from without, and a personal confession

Imagine a future where scientists depend on AI to refine and enhance their own writing, and even entrust it with drafting entire sections of their manuscripts. Where AI assists in summarizing relevant literature, identifying gaps, and uncovering blind spots. Where AI aids in data analysis, hypothesis generation, and experiment design. And where AI takes on the role of human reviewers, by evaluating journal submissions and grant proposals.

The first lines of this book were written at a time when all of this still felt like a distant future. OpenAI's ChatGPT had only recently been released, and the stir it had caused had yet to reach the shores of mainland Europe—or so I make myself believe. Like many others, I was still sceptical about AI's potential. No more than two(!) years later, there is no doubt that the future I once imagined has already materialized. It is therefore only natural to briefly reflect on the (potential) impact of AI on science's poly-crisis.

Let's start with the bad news: apart from the fact that AI research may be going through a replication crisis itself,[1] AI may worsen science's poly-crisis by expanding the toolkit of fraudsters and researchers who want to cut corners. To give just a couple of examples: AI facilitates fabricating realistic but entirely fictitious data,[2] and setting up paper mills (i.e., businesses that fabricate and sell scholarly manuscripts);[3] it makes

1 See Kapoor & Narayanan, 2023.
2 See Chen et al., 2024.
3 See Hosseini et al., 2025.

 https://doi.org/10.11647/OBP.0507.07

data-dredging—exhaustively testing hypotheses, and only reporting significant results—particularly easy; and, to achieve desired results, researchers may leverage existing algorithmic biases, or intentionally introduce new ones by feeding the system biased training data.

Further, AI poses challenges to transparency. For one, analyses are often challenging to replicate due to the black-box nature of widely used AI algorithms, particularly proprietary ones. For another, researchers may be reluctant to share their materials for fear of unacknowledged appropriation by (commercial) AI tools.

Finally, AI accelerates the discovery process—e.g., among the estimated ten^{60} drug-like possible molecules, AI can be trained to identify the most promising ones, and thus help to prioritize experimental investigation.[4] This is good in itself. However, with discovery becoming easier and current incentives pushing for novelty, there is a significant risk that the scientific literature will be biased—even more than it currently is—towards new findings that, once published, remain poorly tested.

How to respond to these challenges? A first thing to note is that they are not fundamentally different from those discussed at the start of this book—they may intensify the poly-crisis but do not alter its nature. That is, AI facilitates but is not the first cause of questionable research practices, lack of transparency, and the proliferation of inadequately tested new ideas. In this sense, there is no need for proposals tailored specifically to AI—AI just amplifies the urgency to implement those proposals from the previous chapter.

Let me make this a bit more explicit. In Chapter 2, I argued that a principal cause of the poly-crisis is science's current incentive structure, which heavily pressures researchers to produce not just positive, but positive *novel* results. The scientific coordination outlined in Chapter 6 promotes a more balanced distribution of rewards—research that produces negative results or builds on existing ideas is as likely to earn recognition and funding as research that delivers positive, novel findings. In making this so, scientific coordination reduces the appeal of rushed research (whether or not it involves AI), secrecy (whether or not it involves AI), and obsessive novelty-seeking (whether or not it involves AI).

4 See Wang et al., 2023.

To be clear, scientific coordination alone will not resolve the poly-crisis. Greater investment in research integrity education, responsible research guidelines, AI usage standards, regulatory frameworks, and related efforts may still be necessary.[5] However, when it comes to science's incentive structure, AI does not appear to introduce additional challenges.

And there is more comforting news: AI can potentially be harnessed to help address the poly-crisis. For example, AI's core strength in detecting patterns within large datasets has proven valuable in evaluating the reproducibility of scientific papers. Leveraging this capability, Yang and colleagues developed an AI tool that predicts a study's likelihood of replication success with approximately 75% accuracy.[6] This tool enables scientists, reviewers, and funders to quickly assess the credibility of research findings. Researchers can also use it to 'stress-test' their studies before submission, providing an opportunity to pause, reconsider their approach, and identify potential corrections.[7] Moreover, AI has enabled the development of a suite of tools to identify questionable research practices—the tools include data integrity checkers, plagiarism detectors, transparency and explainability tools, data provenance assessment tools, and AI model auditing tools.[8] Another tool, called FAIRMetaText, uses AI to analyse and enhance metadata quality, which can help researchers in producing FAIRer datasets.[9]

Additionally, Generative AI's proficiency at summarizing and simplifying complex texts can improve meaningful access to scientific content.[10] This benefits both fellow researchers—especially those from other disciplines—and the wider public, making it easier for them to engage in Peirce-style 2.0 research prioritization and related activities. By aiding authors with text generation, Generative AI also lowers barriers for non-native English-speaking researchers to contribute to science—including RED AREA (see Figure 1.2) work.

Forecasting the future progress of AI technology is difficult. While

---

5 See Chen et al., 2024.
6 See Yang et al., 2020.
7 See https://insight.kellogg.northwestern.edu/article/how-ai-can-help-researchers-navigate-the-replication-crisis; quoted in Straiton, 2024.
8 See Chen et al., 2024.
9 See Sundaram & Musen, 2023.
10 See Hosseini et al., 2025.

I generally tend to be pessimistic, the points above lead me to believe that, provided a commitment to re-organizing science along the lines of Chapter 6, caution is warranted but outright alarm is not.

Arguably a much bigger cause for concern is a crisis from without: the attack on science by the likes of Viktor Orbán (Hungary), Geert Wilders (the Netherlands), and Donald Trump (U.S.). Viktor Orbán transferred control of the Hungarian Academy of Sciences to a government-run body, enabling his political allies to influence research funding and priorities. He also handed over major public universities to private organizations led by his associates, banned gender studies, interfered in climate change discourse, and cut funding for politically sensitive research areas, including migration, LGBTQ+ rights, and corruption. In the Netherlands, the demissionary government—which Geert Wilders' far-right Party for Freedom (PVV) is a part of—has implemented a €1 billion budget cut for higher education, along with measures to reduce the number of international students at Dutch universities. In the U.S., the Silencing Science Tracker recorded 346 anti-science actions during Trump's first administration, including censorship, budget cuts, personnel changes, research obstruction, and misrepresentation of scientific findings.[11] The second Trump administration is planning/executing further cuts, targeting the National Institutes of Health (NIH), the National Science Foundation (NSF), and the National Oceanic and Atmospheric Administration (NOAA). Additional measures include defunding universities, reducing student loan programmes, downsizing and restructuring scientific agencies, dismantling climate initiatives, and exerting greater control over research priorities through the re-establishment of the President's Council on Science and Technology. And also elsewhere, far-right parties have been effective in stoking anti-science sentiments, including in France, Italy, Belgium, Austria, the Czech Republic, Poland, and Slovakia.

In this climate, my emphasis on the crisis in science could inadvertently provide authoritarian parties with justification to further defund science. My proposals for reform risk misinterpretation too—the far-right could seize upon my call for coordination to justify centralized control, allowing it to dictate not only what scientists should study but

11 See Webb & Kurtz, 2022.

even how and what they should think.

I can only approach this with humility—I genuinely don't know how to prevent such appropriation. Yet, the risk of misuse cannot be a reason to remain silent. I care too deeply about science to accept its current state, and I refuse to self-censor—doing so would mean they have already won, and in fact would compromise my scientific integrity.

I began the introductory chapter with a personal note. I shared that, for a couple of years, I had been an obsessive neomaniac, that my constant pursuit of the new eventually led to innovation fatigue, and that that condition became a key motivation for writing this book. But along the way, I have come to realize that the very story I have told is infused with novelty: I couldn't resist offering a novel explanation for science's poly-crisis, nor have I held back from pushing myself to develop new solutions for it.

I could easily shift the blame onto my editor—or book editors in general—arguing that they simply don't publish old news. But that would be dishonest. The truth is, I have once again felt the deep satisfaction of, as Bacon put it, going 'off the beaten track'. And that may be the biggest shortcoming of my account: reforming science's incentive structure may prove ineffective if science remains exclusively populated by individuals who love the new—who have internalized what Wisnioski calls society's innovation imperative.[12] In other words, as I noted in Chapter 6, we may also need to prioritize attracting individuals who are *intrinsically* motivated to perform essential but routine tasks, such as replication and other RED AREA work. But because the Science of Team Science is still in its early stages, we lack robust evidence-based strategies for effectively identifying, selecting, and rewarding such individuals, as well as for fostering environments in which they can thrive.

So if you are a researcher and eager to go 'off the beaten track' yourself, that is at least one area you could go.

12 See Wisnioski, 2025.

Acknowledgements

Many of the ideas for this book stem from conversations with members of META/e, whose input was invaluable throughout the process. Among them, two people stand out: Sajedeh Rasti and Daniel Lakens, my scientific-coordination-buddies, whose insights and help with some of the more technical parts of the book were especially important.

I am deeply indebted to those who read and commented on earlier versions of this manuscript (in order of appearance): Elizabeth O'Neill, Vlasta Sikimić, Wybo Houkes, and Remy Hertogs. I also appreciate the two official reviewers, Thomas Reydon and an unusually constructive Reviewer 2—if all peer review were this generous and timely, we'd all publish faster and better.

Gratitude goes to Alessandra Tosi, my editor, and Annie Hine, my copy-editor, for their diligence and enthusiasm, and to Open Book Publishers for making it possible to publish this book in the spirit of the principles it champions—true Diamond Open Access.

I would like to acknowledge MS Copilot for assisting me with text-editing the revisions I made in response to the reviewers' suggestions and for helping craft the abstracts of this book.

Finally, heartfelt appreciation to Marjolijn, Briek, and Claire for knowing exactly how to pull me through bouts of innovation fatigue.

References

Aagaard, K., Kladakis, A., & Nielsen, M. W. (2020). Concentration or dispersal of research funding? *Quantitative Science Studies*, *1*(1), 117–149. https://doi.org/10.1162/qss_a_00002

Abritis, A., McCook, A., & RETRACTION WATCH. (2017). Cash incentives for papers go global. *Science*, *357*(6351), 541–541. https://doi.org/10.1126/science.357.6351.541

Aldieri, L., Bruno, B., & Vinci, C. P. (2021). A multi-dimensional approach to happiness and innovation. *Applied Economics*, *53*(11), 1300–1310. https://doi.org/10.1080/00036846.2020.1828807

Anadon, L. D., Chan, G., Harley, A. G., Matus, K., Moon, S., Murthy, S. L., & Clark, W. C. (2016). Making technological innovation work for sustainable development. *Proceedings of the National Academy of Sciences*, *113*(35), 9682–9690. https://doi.org/10.1073/pnas.1525004113

Anvari, F., & Lakens, D. (2018). The replicability crisis and public trust in psychological science. *Comprehensive Results in Social Psychology*, *3*(3), 266–286. https://doi.org/10.1080/23743603.2019.1684822

Arendt, H. (1958). *The Human Condition*. University of Chicago Press.

Arrow, K. J. (1959). *Economic Welfare and the Allocation of Resources for Invention*. The Rand Corporation.

Arrowsmith, J. (2011). Phase II failures: 2008–2010. *Nature Reviews Drug Discovery*, *10*(5), Article 5. https://doi.org/10.1038/nrd3439

Asrar, G. R., Hurrell, J. W., & Busalacchi, A. J. (2013). The World Climate Research Program Strategy and Priorities: Next Decade. In G. R. Asrar & J. W. Hurrell (Eds), *Climate Science for Serving Society: Research, Modeling and Prediction Priorities* (pp. 1–12). Springer Netherlands. https://doi.org/10.1007/978-94-007-6692-1_1

Bacon, F. (1620/1949). *Novum Organum* (P. Urbach & J. Gibson, Trans.). P.F. Collier.

Bacon, F. (1624/1857). De Dignitate et Augmentis Scientiarum. In J. Spedding, R. L. Ellis, & D. D. Heath (Eds), *The Works of Francis Bacon* (Vol. IV) (pp. 275–438). Longmans & Co.

Baker, M. (2016). 1,500 scientists lift the lid on reproducibility. *Nature*, *533*(7604), Article 7604. https://doi.org/10.1038/533452a

Banks, G. C., Field, J. G., Oswald, F. L., O'Boyle, E. H., Landis, R. S., Rupp, D. E., & Rogelberg, S. G. (2019). Answers to 18 Questions About open science Practices. *Journal of Business and Psychology*, *34*(3), 257–270. https://doi.org/10.1007/s10869-018-9547-8

Bargh, J. A., Chen, M., & Burrows, L. (1996). Automaticity of social behavior: Direct effects of trait construct and stereotype activation on action. *Journal of Personality and Social Psychology*, *71*(2), 230–244. https://doi.org/10.1037/0022-3514.71.2.230

Bausell, R. B. (2021). *The Problem with Science: The Reproducibility Crisis and What to Do About It*. Oxford University Press. https://doi.org/10.1093/oso/9780197536537.001.0001

Begley, C. G., & Ellis, L. M. (2012). Raise standards for preclinical cancer research. *Nature*, *483*(7391), Article 7391. https://doi.org/10.1038/483531a

Bem, D. J. (2011). Feeling the future: Experimental evidence for anomalous retroactive influences on cognition and affect. *Journal of Personality and Social Psychology*, *100*(3), 407–425. https://doi.org/10.1037/a0021524

Bergh, D. D., Sharp, B. M., Aguinis, H., & Li, M. (2017). Is there a credibility crisis in strategic management research? Evidence on the reproducibility of study findings. *Strategic Organization*, *15*(3), 423–436. https://doi.org/10.1177/1476127017701076

Berkman, E. T., & Wilson, S. M. (2021). So Useful as a Good Theory? The Practicality Crisis in (Social) Psychological Theory. *Perspectives on Psychological Science*, *16*(4), 864–874. https://doi.org/10.1177/1745691620969650

Bigg, C., Reinisch, J., Somsen, G., & Widmalm, S. (2023). The art of gathering: Histories of international scientific conferences. *The British Journal for the History of Science*, *56*(4), 423–433. https://doi.org/10.1017/S0007087423000638

Biggi, G., & Giuliani, E. (2020). The noxious consequences of innovation: What do we know? *Industry and Innovation*, *28*(1), 19–41. https://doi.org/10.1080/13662716.2020.1726729

Botvinik-Nezer, R., Holzmeister, F., Camerer, C. F., Dreber, A., Huber, J., Johannesson, M., Kirchler, M., Iwanir, R., Mumford, J. A., Adcock, R. A., Avesani, P., Baczkowski, B. M., Bajracharya, A., Bakst, L., Ball, S., Barilari, M., Bault, N., Beaton, D., Beitner, J., … Schonberg, T. (2020). Variability in the analysis of a single neuroimaging dataset by many teams. *Nature*, *582*(7810), 84–88. https://doi.org/10.1038/s41586-020-2314-9

Boudreau, K. J., Guinan, E. C., Lakhani, K. R., & Riedl, C. (2016). Looking Across and Looking Beyond the Knowledge Frontier: Intellectual Distance, Novelty, and Resource Allocation in Science. *Management Science*, *62*(10), 2765–2783. https://doi.org/10.1287/mnsc.2015.2285

Boutroux, P. (1914). *Les principes de l'analyse mathématique*. Hermann.

Bratton, L., Adams, R. C., Challenger, A., Boivin, J., Bott, L., Chambers, C. D., & Sumner, P. (2019). The association between exaggeration in health-related science news and academic press releases: A replication study. *Wellcome Open Research*, *4*, 148. https://doi.org/10.12688/wellcomeopenres.15486.2

Breznau, N., Rinke, E. M., Wuttke, A., Nguyen, H. H. V., Adem, M., Adriaans, J., Alvarez-Benjumea, A., Andersen, H. K., Auer, D., Azevedo, F., Bahnsen, O., Balzer, D., Bauer, G., Bauer, P. C., Baumann, M., Baute, S., Benoit, V., Bernauer, J., Berning, C., ... Żółtak, T. (2022). Observing many researchers using the same data and hypothesis reveals a hidden universe of uncertainty. *Proceedings of the National Academy of Sciences*, *119*(44), e2203150119. https://doi.org/10.1073/pnas.2203150119

Brodeur, A., Esterling, K., Ankel-Peters, J., Bueno, N. S., Desposato, S., Dreber, A., Genovese, F., Green, D. P., Hepplewhite, M., Hoces de la Guardia, F., Johannesson, M., Kotsadam, A., Miguel, E., Velez, Y. R., & Young, L. (2024). Promoting Reproducibility and Replicability in Political Science. *Research & Politics*, *11*(1), 20531680241233439. https://doi.org/10.1177/20531680241233439

Bruns, S. B., Deressa, T. K., Stanley, T. D., Doucouliagos, C., & Ioannidis, J. P. A. (2024). Estimating the extent of selective reporting: An application to economics. *Research Synthesis Methods*, *15*(4), 590–602. https://doi.org/10.1002/jrsm.1711

Brunschvicg, L. (1912). *Les étapes de la philosophie mathématique*. Alcan.

Brynjolfsson, E., & McAfee, A. (2012). *Race Against the Machine: How the Digital Revolution is Accelerating Innovation, Driving Productivity, and Irreversibly Transforming Employment and the Economy*. Digital Frontier Press. 3

Busalacchi, A. J., & Asrar, G. R. (2009). World Climate Research Programme: Achievements, activities and challenges. *WMO Bulletin*, *58*(3), 151–161.

Bush, V. (1945). *Science, the Endless Frontier: A Report to the President*. United States Government Printing Office. https://www.nsf.gov/od/lpa/nsf50/vbush1945.htm

Butler, L.-A., Matthias, L., Simard, M.-A., Mongeon, P., & Haustein, S. (2023). The oligopoly's shift to open access: How the big five academic publishers profit from article processing charges. *Quantitative Science Studies*, *4*(4), 778–799. https://doi.org/10.1162/qss_a_00272

Button, K. S., Ioannidis, J. P. A., Mokrysz, C., Nosek, B. A., Flint, J., Robinson, E. S. J., & Munafò, M. R. (2013). Power failure: Why small sample size undermines the reliability of neuroscience. *Nature Reviews Neuroscience*, *14*(5), 365–376. https://doi.org/10.1038/nrn3475

Camerer, C. F., Dreber, A., Forsell, E., Ho, T.-H., Huber, J., Johannesson, M., Kirchler, M., Almenberg, J., Altmejd, A., Chan, T., Heikensten, E., Holzmeister, F., Imai, T., Isaksson, S., Nave, G., Pfeiffer, T., Razen, M., & Wu,

H. (2016). Evaluating replicability of laboratory experiments in economics. *Science*, *351*(6280), 1433–1436. https://doi.org/10.1126/science.aaf0918

Camerer, C. F., Dreber, A., Holzmeister, F., Ho, T.-H., Huber, J., Johannesson, M., Kirchler, M., Nave, G., Nosek, B. A., Pfeiffer, T., Altmejd, A., Buttrick, N., Chan, T., Chen, Y., Forsell, E., Gampa, A., Heikensten, E., Hummer, L., Imai, T., ... Wu, H. (2018). Evaluating the replicability of social science experiments in Nature and Science between 2010 and 2015. *Nature Human Behaviour*, 2(9), 637–644. https://doi.org/10.1038/s41562-018-0399-z

Cameron, G. (1998). *Innovation and Growth: A Survey of the Empirical Evidence.* Working Paper. Nuffield College, Oxford University.

Candolle, A. de. (1873). *Histoire des sciences et des savants depuis deux siècles: Précédée et suivie d'autres études sur les sujets scientifiques, en particluier sur l'hérédité et la sélection dans l'espèce humaine.* H. Georg.

Case, C. M. (1928). Scholarship in sociology. *Sociology and Social Research, 12,* 323–340.

Chalmers, I., & Glasziou, P. (2009). Avoidable waste in the production and reporting of research evidence. *The Lancet*, *374*(9683), 86–89. https://doi.org/10.1016/S0140-6736(09)60329-9

Chang, A. C., & Li, P. (2015). Is Economics Research Replicable? Sixty Published Papers from Thirteen Journals Say "Usually Not." *Finance and Economics Discussion Series.* https://www.federalreserve.gov/econres/feds/is-economics-research-replicable-sixty-published-papers-from-thirteen-journals-say-quotusually-notquot.htm

Chen, S. (2024). A wave of retractions is shaking physics. *MIT Technology Review.* https://www.technologyreview.com/2024/05/15/1092535/a-wave-of-retractions-is-shaking-physics/

Chen, Z., Chen, C., Yang, G., He, X., Chi, X., Zeng, Z., & Chen, X. (2024). Research integrity in the era of artificial intelligence: Challenges and responses. *Medicine*, *103*(27), e38811. https://doi.org/10.1097/MD.0000000000038811

Clark, T. D., Raby, G. D., Roche, D. G., Binning, S. A., Speers-Roesch, B., Jutfelt, F., & Sundin, J. (2020). Ocean acidification does not impair the behaviour of coral reef fishes. *Nature*, *577*(7790), 370–375. https://doi.org/10.1038/s41586-019-1903-y

Coady, S. A., Mensah, G. A., Wagner, E. L., Goldfarb, M. E., Hitchcock, D. M., & Giffen, C. A. (2017). Use of the National Heart, Lung, and Blood Institute Data Repository. *New England Journal of Medicine*, *376*(19), 1849–1858. https://doi.org/10.1056/NEJMsa1603542

Cohen, B. A. (2017). How should novelty be valued in science? *eLife*, *6*, e28699. https://doi.org/10.7554/eLife.28699

Cohen, M. R. (1926). The Myth about Bacon and the Inductive Method. *The Scientific Monthly*, *23*(6), 504–508.

Coles, N. A., Hamlin, J. K., Sullivan, L. L., Parker, T. H., & Altschul, D. (2022a). Build up big-team science. *Nature, 601*(7894), 505–507. https://doi.org/10.1038/d41586-022-00150-2

Coles, N. A., March, D. S., Marmolejo-Ramos, F., ... Liuzza, M. T. (2022b). A multi-lab test of the facial feedback hypothesis by the Many Smiles Collaboration. *Nature Human Behaviour, 6*, 1731–1742. https://doi.org/10.1038/s41562-022-01458-9

Conway, P. (2018). Another Science Is Possible By Isabelle Stengers. *Society and Space.* https://www.societyandspace.org/articles/another-science-is-possible-by-isabelle-stengers

Coretta, S., Casillas, J. V., Roessig, S., Franke, M., Ahn, B., Al-Hoorie, A. H., Al-Tamimi, J., Alotaibi, N. E., AlShakhori, M. K., Altmiller, R. M., Arantes, P., Athanasopoulou, A., Baese-Berk, M. M., Bailey, G., Sangma, C. B. A., Beier, E. J., Benavides, G. M., Benker, N., BensonMeyer, E. P., ... Roettger, T. B. (2023). Multidimensional Signals and Analytic Flexibility: Estimating Degrees of Freedom in Human-Speech Analyses. *Advances in Methods and Practices in Psychological Science, 6*(3), 25152459231162567. https://doi.org/10.1177/25152459231162567

Daitch, V., Adi Turjeman, Poran, I., Tau, N., Ayalon-Dangur, I., Nashashibi, J., Yahav, D., Paul, M., & Leibovici, L. (2022). Underrepresentation of women in randomized controlled trials: A systematic review and meta-analysis. *Trials, 23*(1), 1038. https://doi.org/10.1186/s13063-022-07004-2

Dempster, G., Sutherland, G., & Keogh, L. (2022). Scientific research in news media: A case study of misrepresentation, sensationalism and harmful recommendations. *Journal of Science Communication, 21*(1), A06. https://doi.org/10.22323/2.21010206

Descartes, R. (1637/1899). *A Discourse on the Method* (J. Veitch, Trans.). Open Court.

Douglas, H. (2005). Inserting the Public Into Science. In S. Maasen & P. Weingart (Eds), *Democratization of Expertise? Exploring Novel Forms of Scientific Advice in Political Decision-Making* (pp. 153–169). Springer Netherlands. https://doi.org/10.1007/1-4020-3754-6_9

Doyen, S., Klein, O., Pichon, C.-L., & Cleeremans, A. (2012). Behavioral Priming: It's All in the Mind, but Whose Mind? *PLOS ONE, 7*(1), e29081. https://doi.org/10.1371/journal.pone.0029081

Duhem, P. (1905). *Les Origines de la statique*. Hermann.

Duhem, P. (1906). *La théorie physique: Son object et sa structure.* Chevalier & Rivière.

Ebersole, C. R., Atherton, O. E., Belanger, A. L., Skulborstad, H. M., Allen, J. M., Banks, J. B., Baranski, E., Bernstein, M. J., Bonfiglio, D. B. V., Boucher, L., Brown, E. R., Budiman, N. I., Cairo, A. H., Capaldi, C. A., Chartier, C. R., Chung, J. M., Cicero, D. C., Coleman, J. A., Conway, J. G., ... Nosek, B.

A. (2016). Many Labs 3: Evaluating participant pool quality across the academic semester via replication. *Journal of Experimental Social Psychology, 67*, 68–82. https://doi.org/10.1016/j.jesp.2015.10.012

Ebersole, C. R., Mathur, M. B., Baranski, E., Bart-Plange, D.-J., Buttrick, N. R., Chartier, C. R., Corker, K. S., Corley, M., Hartshorne, J. K., IJzerman, H., Lazarević, L. B., Rabagliati, H., Ropovik, I., Aczel, B., Aeschbach, L. F., Andrighetto, L., Arnal, J. D., Arrow, H., Babincak, P., ... Nosek, B. A. (2020). Many Labs 5: Testing Pre-Data-Collection Peer Review as an Intervention to Increase Replicability. *Advances in Methods and Practices in Psychological Science, 3*(3), 309–331. https://doi.org/10.1177/2515245920958687

Edgerton, D. (2006). *The Shock of the Old: Technology and Global History Since 1900*. Profile Books.

Eronen, M. I., & Bringmann, L. F. (2021). The Theory Crisis in Psychology: How to Move Forward. *Perspectives on Psychological Science, 16*(4), 779–788. https://doi.org/10.1177/1745691620970586

Errington, T. M., Mathur, M., Soderberg, C. K., Denis, A., Perfito, N., Iorns, E., & Nosek, B. A. (2021). Investigating the replicability of preclinical cancer biology. *eLife, 10*, e71601. https://doi.org/10.7554/eLife.71601

Etzkowitz, H. (2002). *MIT and the Rise of Entrepreneurial Science*. Routledge. https://doi.org/10.4324/9780203216675

Etzkowitz, H. (2003). Research groups as 'quasi-firms': The invention of the entrepreneurial university. *Research Policy, 32*(1), 109–121. https://doi.org/10.1016/S0048-7333(02)00009-4

Etzkowitz, H. (2008). *The Triple Helix: University-Industry-Government Innovation in Action*. Routledge. https://doi.org/10.4324/9780203929605

Evans, J. H. (2014). Faith in science in global perspective: Implications for transhumanism. *Public Understanding of Science, 23*(7), 814–832. https://doi.org/10.1177/0963662514523712

Fang, F. C., & Casadevall, A. (2011). Retracted Science and the Retraction Index. *Infection and Immunity, 79*(10), 3855–3859. https://doi.org/10.1128/IAI.05661-11

Farrington, B. (1949). *Francis Bacon: Philosopher of Industrial Science*. Schuman.

Felt, U. (2015). Social Science Expertise in European Policy Innovation Policy. In *Future Directions for Scientific Advice in Europe* (pp. 113–122). Centre for Science and Policy.

Ferrari Dacrema, M., Cremonesi, P., & Jannach, D. (2019). Are we really making much progress? A worrying analysis of recent neural recommendation approaches. *Proceedings of the 13th ACM Conference on Recommender Systems*, 101–109. https://doi.org/10.1145/3298689.3347058

Ferraro, P. J., & Shukla, P. (2020). Feature—Is a Replicability Crisis on the Horizon for Environmental and Resource Economics? *Review of Environmental*

Economics and Policy, 14(2), 339–351. https://doi.org/10.1093/reep/reaa011

Ford, M. (2016). *Rise of the Robots: Technology and the Threat of a Jobless Future* (First paperback edition). Basic Books.

Forscher, P. S., Wagenmakers, E.-J., Coles, N. A., Silan, M. A., Dutra, N., Basnight-Brown, D., & IJzerman, H. (2023). The Benefits, Barriers, and Risks of Big-Team Science. *Perspectives on Psychological Science, 18*(3), 607–623. https://doi.org/10.1177/17456916221082970

Fragkandreas, T. (2022). Three decades of research on innovation and inequality: Causal scenarios, explanatory factors and suggestions. *Prometheus, 38*(2), 147–193. DOI:10.13169/prometheus.38.2.0147

Fraley, R. C., Chong, J. Y., Baacke, K. A., Greco, A. J., Guan, H., & Vazire, S. (2022). Journal N-Pact Factors From 2011 to 2019: Evaluating the Quality of Social/Personality Journals With Respect to Sample Size and Statistical Power. *Advances in Methods and Practices in Psychological Science, 5*(4), 25152459221120217. https://doi.org/10.1177/25152459221120217

Freilich, J. (2020). The Replicability Crisis in Patent Law. *95 Indiana Law Journal 431 (2020), 95*(2). https://www.repository.law.indiana.edu/ilj/vol95/iss2/3

Frias-Navarro, D., Pascual-Llobell, J., Pascual-Soler, M., Perezgonzalez, J., & Berrios-Riquelme, J. (2020). Replication crisis or an opportunity to improve scientific production? *European Journal of Education, 55*(4), 618–631. https://doi.org/10.1111/ejed.12417

Fried, B. J., Boers, M., & Baker, P. R. (1993). A method for achieving consensus on rheumatoid arthritis outcome measures: The OMERACT conference process. *The Journal of Rheumatology, 20*(3), 548–551.

Frith, U. (2020). Fast Lane to Slow Science. *Trends in Cognitive Sciences, 24*(1), 1–2. https://doi.org/10.1016/j.tics.2019.10.007

Fuller, S. (2009). The Genealogy of Judgement: Towards a Deep History of Academic Freedom. *British Journal of Educational Studies, 57*(2), 164–177. https://doi.org/10.1111/j.1467-8527.2009.00434.x

Gabriel, A. S., & Wessel, J. L. (2013). A Step Too Far? Why Publishing Raw Datasets May Hinder Data Collection. *Industrial and Organizational Psychology, 6*(3), 287–290. https://doi.org/10.1111/iops.12051

Galison, P., & Daston, L. (2008). Scientific Coordination as Ethos and Epistemology. In *Instruments in Art and Science* (pp. 296–333). De Gruyter.

Gandal, N., Shur-Ofry, M., Crystal, M., & Shilony, R. (2021). Out of sight: Patents that have never been cited. *Scientometrics, 126*(4), 2903–2929. https://doi.org/10.1007/s11192-020-03849-z

Garisto, D. (2023). *Road Map for U.S. Particle Physics Wins Broad Approval.* Scientific American. https://www.scientificamerican.com/article/road-map-for-u-s-particle-physics-wins-broad-approval/

Gaskell, A. (2019). Technology Isn't Destroying Jobs, But Is Increasing Inequality. *Forbes Magazine*. https://www.forbes.com/sites/adigaskell/2019/05/03/technology-isnt-destroying-jobs-but-is-increasing-inequality/?sh=8aa20005e784

Gibbons, M. (1999). Science's new social contract with society. *Nature*, *402*(6761), C81–C84. https://doi.org/10.1038/35011576

Giner-Sorolla, R. (2019). From crisis of evidence to a "crisis" of relevance? Incentive-based answers for social psychology's perennial relevance worries. *European Review of Social Psychology*, *30*(1), 1–38. https://doi.org/10.1080/10463283.2018.1542902

Godin, B. (2015). *Innovation Contested: The Idea of Innovation Over the Centuries*. Routledge.

Godin, B. (2016a). Making sense of innovation: From weapon to instrument to buzzword. *Quaderni. Communication, Technologies, Pouvoir*, *90*, Article 90. https://doi.org/10.4000/quaderni.977

Godin, B. (2016b). *Representation of Innovation in Seventeenth-Century England*. https://doi.org/10.3167/choc.2016.110202

Godin, B. (2017). *Models of Innovation: The History of an Idea*. The MIT Press. https://doi.org/10.7551/mitpress/10782.001.0001

Godin, B., & Vinck, D. (2017). *Critical Studies of Innovation: Alternative Approaches to the Pro-Innovation Bias*. Edward Elgar Publishing.

Goldfarb, R. S. (1997). Now you see it, now you don't: Emerging contrary results in economics. *Journal of Economic Methodology*, *4*(2), 221–244. https://doi.org/10.1080/13501789700000016

Gordon, M., Viganola, D., Bishop, M., Chen, Y., Dreber, A., Goldfedder, B., Holzmeister, F., Johannesson, M., Liu, Y., Twardy, C., Wang, J., & Pfeiffer, T. (2020). Are replication rates the same across academic fields? Community forecasts from the DARPA SCORE programme. *Royal Society open science*, *7*(7), 200566. https://doi.org/10.1098/rsos.200566

Gould, E., Fraser, H. S., Parker, T. H., Nakagawa, S., Griffith, S. C., Vesk, P. A., Fidler, F., Hamilton, D. G., Abbey-Lee, R. N., Abbott, J. K., Aguirre, L. A., Alcaraz, C., Aloni, I., Altschul, D., Arekar, K., Atkins, J. W., Atkinson, J., Baker, C. M., Barrett, M., … Whelan, S. (2025). Same data, different analysts: Variation in effect sizes due to analytical decisions in ecology and evolutionary biology. *BMC Biology*, *23*(1), 35. https://doi.org/10.1186/s12915-024-02101-x

Graeber, D. (2019). *Bullshit Jobs* (First Simon&Schuster trade paperback edition). Simon & Schuster Paperbacks.

Graudal, N. A., Hubeck-Graudal, T., & Jurgens, G. (2011). Effects of low sodium diet versus high sodium diet on blood pressure, renin, aldosterone, catecholamines, cholesterol, and triglyceride. *Cochrane Database of Systematic Reviews*, *11*. https://doi.org/10.1002/14651858.CD004022.pub3

Guzzo, R. A., Schneider, B., & Nalbantian, H. R. (2022). Open science, closed doors: The perils and potential of open science for research in practice. *Industrial and Organizational Psychology, 15*(4), 495–515. https://doi.org/10.1017/iop.2022.61

Hall, K. L., Vogel, A. L., Huang, G. C., Serrano, K. J., Rice, E. L., Tsakraklides, S. P., & Fiore, S. M. (2018). The science of team science: A review of the empirical evidence and research gaps on collaboration in science. *American Psychologist, 73*(4), 532–548. https://doi.org/10.1037/amp0000319

Hartl, P. (2012). Michael Polanyi on Freedom of Science. *Synthesis Philosophica, 54*, 307–321.

Hazelkorn, E. (2015). *Rankings and the Reshaping of Higher Education*. Palgrave Macmillan U.K. https://doi.org/10.1057/9781137446671

Henderson, R., Jaffe, A. B., & Trajtenberg, M. (1998). Universities as a Source of Commercial Technology: A Detailed Analysis of University Patenting, 1965-1988. *The Review of Economics and Statistics, 80*(1), 119–127.

Hendriks, F., Kienhues, D., & Bromme, R. (2015). Measuring Laypeople's Trust in Experts in a Digital Age: The Muenster Epistemic Trustworthiness Inventory (METI). *PLOS ONE, 10*(10), e0139309. https://doi.org/10.1371/journal.pone.0139309

Henrich, J., Heine, S. J., & Norenzayan, A. (2010). The weirdest people in the world? *Behavioral and Brain Sciences, 33*(2–3), 61–83. https://doi.org/10.1017/S0140525X0999152X

Hensel, P. G. (2021). Reproducibility and replicability crisis: How management compares to psychology and economics – A systematic review of literature. *European Management Journal, 39*(5), 577–594. https://doi.org/10.1016/j.emj.2021.01.002

Herbert, D. L., Barnett, A. G., & Graves, N. (2013). Australia's grant system wastes time. *Nature, 495*(7441), 314–314. https://doi.org/10.1038/495314d

Hewlett, S., Wit, M. de, Richards, P., Quest, E., Hughes, R., Heiberg, T., & Kirwan, J. (2006). Patients and professionals as research partners: Challenges, practicalities, and benefits. *Arthritis and Rheumatism, 55*(4), 676–680. https://doi.org/10.1002/art.22091

Hoekstra, R., & Vazire, S. (2021). Aspiring to greater intellectual humility in science. *Nature Human Behaviour, 5*(12), Article 12. https://doi.org/10.1038/s41562-021-01203-8

Hofelich Mohr, A., Carlson, J., Lizhao, G., Herndon, J., Kozlowski, W., Moore, J., Petters, J., Taylor, S., & Hudson Vitale, C. (2024). *Making Research Data Publicly Accessible: Estimates of Institutional & Researcher Expenses*. Association of Research Libraries. https://doi.org/10.29242/report.radsexpense2024.

Hofstra, B., Kulkarni, V. V., Munoz-Najar Galvez, S., He, B., Jurafsky, D., & McFarland, D. A. (2020). The Diversity–Innovation Paradox in Science.

Proceedings of the National Academy of Sciences, 117(17), 9284–9291. https://doi.org/10.1073/pnas.1915378117

Hood, L., & Rowen, L. (2013). The Human Genome Project: Big science transforms biology and medicine. *Genome Medicine, 5*(9), 79. https://doi.org/10.1186/gm483

Hosseini, M., Horbach, S. P. J. M., Holmes, K., & Ross-Hellauer, T. (2025). Open science at the generative AI turn: An exploratory analysis of challenges and opportunities. *Quantitative Science Studies, 6*, 22–45. https://doi.org/10.1162/qss_a_00337

Hostler, T. J. (2023). The Invisible Workload of Open Research. *Journal of Trial & Error*. https://doi.org/10.36850/mr5

Hötte, K., Somers, M., & Theodorakopoulos, A. (2023). Technology and jobs: A systematic literature review. *Technological Forecasting and Social Change, 194*, 122750. https://doi.org/10.1016/j.techfore.2023.122750

Hug, S. E., & Aeschbach, M. (2020). Criteria for assessing grant applications: A systematic review. *Palgrave Communications, 6*(1), Article 1. https://doi.org/10.1057/s41599-020-0412-9

Huntington-Klein, N., Arenas, A., Beam, E., Bertoni, M., Bloem, J. R., Burli, P., Chen, N., Grieco, P., Ekpe, G., Pugatch, T., Saavedra, M., & Stopnitzky, Y. (2021). The influence of hidden researcher decisions in applied microeconomics. *Economic Inquiry, 59*(3), 944–960. https://doi.org/10.1111/ecin.12992

Ioannidis, J. P. A. (2005). Why Most Published Research Findings Are False. *PLOS Medicine, 2*(8), e124. https://doi.org/10.1371/journal.pmed.0020124

Ioannidis, J. P. A. (2008). Why Most Discovered True Associations Are Inflated. *Epidemiology, 19*(5), 640. https://doi.org/10.1097/EDE.0b013e31818131e7

Ioannidis, J. P. A. (2016). Why Most Clinical Research Is Not Useful. *PLOS Medicine, 13*(6), e1002049. https://doi.org/10.1371/journal.pmed.1002049

Ioannidis, J. P. A., Munafò, M. R., Fusar-Poli, P., Nosek, B. A., & David, S. P. (2014). Publication and other reporting biases in cognitive sciences: Detection, prevalence, and prevention. *Trends in Cognitive Sciences, 18*(5), 235–241. https://doi.org/10.1016/j.tics.2014.02.010

IPCC. (2022). *IPCC Sixth Assessment Report*.

Jones, J. W., & Dages, K. D. (2013). A New Era of Large-Scale Data Sharing: A Test Publisher's Perspective. *Industrial and Organizational Psychology, 6*(3), 309–312. https://doi.org/10.1111/iops.12057

Kalleberg, A. L. (2011). *Good Jobs, Bad Jobs: The Rise of Polarized and Precarious Employment Systems in the United States, 1970s to 2000s*. Russell Sage Foundation.

Kapoor, S., & Narayanan, A. (2023). Leakage and the reproducibility crisis in machine-learning-based science. *Patterns*, *4*(9), 100804. https://doi.org/10.1016/j.patter.2023.100804

Kealey, S. N. (2023). Can science get by without your tax money? Just ask them over at IBM. *The Times*. https://www.thetimes.co.uk/article/can-science-get-by-without-your-tax-money-just-ask-them-over-at-ibm-zhdg0kfzmcz

Keraudren, P. (2018). The Contribution of Social Sciences and the Humanities to Research Addressing Societal Challenges. Towards a Policy for Interdisciplinarity at European Level? In A. Tressaud (Ed.), *Progress in Science, Progress in Society* (pp. 43–57). Springer International Publishing. https://doi.org/10.1007/978-3-319-69974-5_5

Kersten, A. (1996). *Een organisatie van en voor onderzoekers: De Nederlandse Organisatie voor Zuiver-Wetenschappelijk Onderzoek (Z.W.O.), 1947-1988*. Van Gorcum.

Kimberly, J. R. (1981). Managerial innovation. In *Handbook of Organizational Design* (pp. 84–104). Oxford University Press.

Kimmel, K., Avolio, M. L., & Ferraro, P. J. (2023). Empirical evidence of widespread exaggeration bias and selective reporting in ecology. *Nature Ecology & Evolution*, *7*(9), 1525–1536. https://doi.org/10.1038/s41559-023-02144-3

Kitcher, P. (2011). *Science in a Democratic Society*. Prometheus Books.

Klein, J., & Giglioni, G. (2020). Francis Bacon. In E. N. Zalta (Ed.), *The Stanford Encyclopedia of Philosophy* (Fall 2020). Metaphysics Research Lab, Stanford University. https://plato.stanford.edu/archives/fall2020/entries/francis-bacon/

Klein, R. A., Cook, C. L., Ebersole, C. R., Vitiello, C., Nosek, B. A., Hilgard, J., Ahn, P. H., Brady, A. J., Chartier, C. R., Christopherson, C. D., Clay, S., Collisson, B., Crawford, J. T., Cromar, R., Gardiner, G., Gosnell, C. L., Grahe, J., Hall, C., Howard, I., ... Ratliff, K. A. (2022). Many Labs 4: Failure to Replicate Mortality Salience Effect With and Without Original Author Involvement. *Collabra: Psychology*, *8*(1), 35271. https://doi.org/10.1525/collabra.35271

Klein, R. A., Vianello, M., Hasselman, F., Adams, B. G., Adams, R. B., Alper, S., Aveyard, M., Axt, J. R., Babalola, M. T., Bahník, Š., Batra, R., Berkics, M., Bernstein, M. J., Berry, D. R., Bialobrzeska, O., Binan, E. D., Bocian, K., Brandt, M. J., Busching, R., ... Nosek, B. A. (2018). Many Labs 2: Investigating Variation in Replicability Across Samples and Settings. *Advances in Methods and Practices in Psychological Science*, *1*(4), 443–490. https://doi.org/10.1177/2515245918810225

Konkol, M., Kray, C., & Pfeiffer, M. (2019). Computational reproducibility in geoscientific papers: Insights from a series of studies with geoscientists and a reproduction study. *International Journal of Geographical Information Science*, *33*(2), 408–429. https://doi.org/10.1080/13658816.2018.1508687

Kotzé, L. J., Kim, R. E., Burdon, P., du Toit, L., Glass, L.-M., Kashwan, P., Liverman, D., Montesano, F. S., Rantala, S., Sénit, C.-A., Treyer, S., &

Calzadilla, P. V. (2022). Planetary Integrity. In C.-A. Sénit, F. Biermann, & T. Hickmann (Eds), *The Political Impact of the Sustainable Development Goals: Transforming Governance Through Global Goals?* (pp. 140–171). Cambridge University Press. https://doi.org/10.1017/9781009082945.007

Kowall, B., Ahrenfeldt, L. J., Basten, J., Becher, H., Brand, T., Braun, J., Casjens, S., Claessen, H., Denz, R., Diebner, H. H., Diexer, S., Eisemann, N., Furrer, E., Galetzka, W., Girschik, C., Karch, A., Mikolajczyk, R., Peters, M., Rospleszcz, S., ... Rübsamen, N. (2025). Marital status and risk of cardiovascular disease – a multi-analyst study in epidemiology. *European Journal of Epidemiology, 40*(5), 497–509. https://doi.org/10.1007/s10654-025-01235-8

Kraemer-Mbula, E., & Wamae, W. (Eds) (2010). *Innovation and the Development Agenda*. OECD.

Krige, J. (2004). I. I. Rabi and the Birth of CERN. *Physics Today, 57*(9), 44–48. https://doi.org/10.1063/1.1809091

Kropp, K. (2021). The EU and the social sciences: A fragile relationship. *The Sociological Review, 69*(6), 1325–1341. https://doi.org/10.1177/00380261211034706

Kropp, K., & Larsen, A. G. (2023). Changing the topics: The social sciences in EU-funded research projects. *Comparative European Politics, 21*(2), 176–207. https://doi.org/10.1057/s41295-022-00313-5

Kuhn, T. (1962). *The Structure of Scientific Revolutions*. University of Chicago Press.

Lakatos, I. (1970). Falsification and the Methodology of Scientific Research Programmes. In I. Lakatos & A. Musgrave (Eds), *Criticism and the Growth of Knowledge* (pp. 91–196). Cambridge University Press.

Lakens, D. (2014). Performing high-powered studies efficiently with sequential analyses. *European Journal of Social Psychology, 44*(7), 701–710. https://doi.org/10.1002/ejsp.2023

Lakens, D. (2020). Pandemic researchers—Recruit your own best critics. *Nature, 581*(7807), 121–121. https://doi.org/10.1038/d41586-020-01392-8

Lakens, D. (2023). *Concerns about Replicability, Theorizing, Applicability, Generalizability, and Methodology across Two Crises in Social Psychology*. OSF. https://doi.org/10.31234/osf.io/dtvs7

LeBel, E. P., & Peters, K. R. (2011). Fearing the Future of Empirical Psychology: Bem's (2011) Evidence of Psi as a Case Study of Deficiencies in Modal Research Practice. *Review of General Psychology, 15*(4), 371–379. https://doi.org/10.1037/a0025172

Leogrande, E., & Nicassio, R. (2021). Collaborative Processes in Science and Literature: An In-Depth Look at the Cases of CERN and SIC. *Frontiers in Research Metrics and Analytics, 5*. https://doi.org/10.3389/frma.2020.592819

Leonelli, S. (2023). *Philosophy of Open Science*. Cambridge University Press.

https://doi.org/10.1017/9781009416368

Leys, R. (2024). *Anatomy of a Train Wreck: The Rise and Fall of Priming Research* (Vol. 1–1 online resource.). University of Chicago Press.

Lösel, F. (2018). Evidence comes by replication, but needs differentiation: The reproducibility issue in science and its relevance for criminology. *Journal of Experimental Criminology, 14*(3), 257–278. https://doi.org/10.1007/s11292-017-9297-z

Luukkonen, T. (2012). Conservatism and risk-taking in peer review: Emerging ERC practices. *Research Evaluation, 21*(1), 48–60. https://doi.org/10.1093/reseval/rvs001

Malmivaara, A. (2019). Generalizability of findings from randomized controlled trials is limited in the leading general medical journals. *Journal of Clinical Epidemiology, 107*, 36–41. https://doi.org/10.1016/j.jclinepi.2018.11.014

Martinez, C., & Sterzi, V. (2019). University patenting and the quest for technology transfer policy models in Europe. In A. Varga & K. Erdos (Eds), *Handbook of Universities and Regional Development*. Edward Elgar Publishing.

Maxwell, L. J., Beaton, D. E., Shea, B. J., Wells, G. A., Boers, M., Grosskleg, S., Bingham, C. O., Conaghan, P. G., D'Agostino, M. A., Wit, M. de, Gossec, L., March, L., Simon, L. S., Singh, J. A., Strand, V., & Tugwell, P. (2019). Core Domain Set Selection According to OMERACT Filter 2.1: The OMERACT Methodology. *The Journal of Rheumatology, 46*(8), 1014–1020. https://doi.org/10.3899/jrheum.181097

Maxwell, N. (1972). A Critique of Popper's Views on Scientific Method. *Philosophy of Science, 39*(2), 131–152.

Mellor, F. (2015). Non-News Values in Science Journalism. In B. Rappert & B. Balmer (Eds), *Absence in Science, Security and Policy* (pp. 93–113). Palgrave Macmillan U.K. https://doi.org/10.1057/9781137493736_5

Merton, R. K. (1957). Priorities in Scientific Discovery: A Chapter in the Sociology of Science. *American Sociological Review, 22*(6), 635–659. https://doi.org/10.2307/2089193

Michie, S., West, R., Campbell, R., Brown, J., & Gainforth, H. (2014). *ABC of Behaviour Change Theories*. Silverback Publishing.

Miłkowski, M., Hensel, W. M., & Hohol, M. (2018). Replicability or reproducibility? On the replication crisis in computational neuroscience and sharing only relevant detail. *Journal of Computational Neuroscience, 45*(3), 163–172. https://doi.org/10.1007/s10827-018-0702-z

Mirowski, P. (2011). *Science-Mart: Privatizing American Science*. Harvard University Press.

Mischel, W. (2008). The Toothbrush Problem. *APS Observer, 21*. https://www.psychologicalscience.org/observer/the-toothbrush-problem

Mitcham, C. (1990). Three ways of being-with technology. In *From Artifact to Habitat: Studies in the Critical Engagement of Technology* (pp. 31–59). LeHigh University Press.

Mortimer, T. (1780). *The Elements of Commerce: Politics and Finances, in Three Treatises on those Important Subjects.* https://historicaltexts.jisc.ac.uk/ecco-0936101100

Nakagawa, S., Armitage, D. W., Froese, T., Yang, Y., & Lagisz, M. (2025). Poor hypotheses and research waste in biology: Learning from a theory crisis in psychology. *BMC Biology, 23*(1), 33. https://doi.org/10.1186/s12915-025-02134-w

Nanay, B. (2011). Popper's Darwinian Analogy. *Perspectives on Science, 19*(3), 337–354. https://doi.org/10.1162/POSC_a_00043

National Academy of Sciences. (2018). Incentives in Science Communication. In *The Science of Science Communication III: Inspiring Novel Collaborations and Building Capacity: Proceedings of a Colloquium*. National Academies Press (U.S.). https://www.ncbi.nlm.nih.gov/books/NBK507368/

Navar, A. M., Pencina, M. J., Rymer, J. A., Louzao, D. M., & Peterson, E. D. (2016). Use of Open Access Platforms for Clinical Trial Data. *JAMA, 315*(12), 1283–1284. https://doi.org/10.1001/jama.2016.2374

Nelkin, D. (1995). *Selling Science: How the Press Covers Science and Technology* (Rev. ed). W.H. Freeman.

Newton, I. (1675). Letter from Sir Isaac Newton to Robert Hooke. Historical Society of Pennsylvania.

North, M. (2015). *Novelty: A History of the New*. The University of Chicago Press.

Nosek, B. A., & Bar-Anan, Y. (2012). Scientific Utopia: I. Opening Scientific Communication. *Psychological Inquiry, 23*(3), 217–243. https://doi.org/10.1080/1047840X.2012.692215

Nosek, B. A., Hardwicke, T. E., Moshontz, H., Allard, A., Corker, K. S., Dreber, A., Fidler, F., Hilgard, J., Struhl, M. K., Nuijten, M. B., Rohrer, J. M., Romero, F., Scheel, A. M., Scherer, L. D., Schönbrodt, F. D., & Vazire, S. (2022). Replicability, Robustness, and Reproducibility in Psychological Science. *Annual Review of Psychology, 73*(Volume 73, 2022), 719–748. https://doi.org/10.1146/annurev-psych-020821-114157

Nosek, B. A., Spies, J. R., & Motyl, M. (2012). Scientific Utopia: II. Restructuring Incentives and Practices to Promote Truth Over Publishability. *Perspectives on Psychological Science, 7*(6), 615–631. https://doi.org/10.1177/1745691612459058

NSF. (1957). *Basic Research: A National Resource*. National Science Foundation [U.S. G.P.O.].

Nye, M. J. (2011). *Michael Polanyi and His Generation: Origins of the Social Construction of Science*. University of Chicago Press.

OPEN SCIENCE COLLABORATION. (2015). Estimating the reproducibility of psychological science. *Science, 349*(6251), aac4716. https://doi.org/10.1126/science.aac4716

Padavic-Callaghan, K. (2024). Physicists are grappling with their own reproducibility crisis. *New Scientist.* https://www.newscientist.com/article/2431927-physicists-are-grappling-with-their-own-reproducibility-crisis/?transaction=1759831129705

Pagell, M. (2021). Replication without repeating ourselves: Addressing the replication crisis in operations and supply chain management research. *Journal of Operations Management, 67*(1), 105–115. https://doi.org/10.1002/joom.1120

Park, M., Leahey, E., & Funk, R. J. (2023). Papers and patents are becoming less disruptive over time. *Nature, 613*(7942), Article 7942. https://doi.org/10.1038/s41586-022-05543-x

Peirce, C. S. (1931). *Collected Papers of Charles Sanders Peirce* (C. Hartshorne, P. Weiss, & A. W. Burks, Eds). The Belknap Press of Harvard University Press.

Pérignon, C., Akmansoy, O., Hurlin, C., Dreber, A., Holzmeister, F., Huber, J., Johannesson, M., Kirchler, M., Menkveld, A. J., Razen, M., & Weitzel, U. (2024). Computational Reproducibility in Finance: Evidence from 1,000 Tests. *The Review of Financial Studies, 37*(11), 3558–3593. https://doi.org/10.1093/rfs/hhae029

Picard, J. (1928). *Essai sur les conditions positives de l'invention dans les sciences.* Alcan.

Pittinsky, T. L. (2015). America's crisis of faith in science. *Science, 348*(6234), 511–512. https://doi.org/10.1126/science.348.6234.511-a

Polanyi, M. (1947). The foundations of academic freedom. *The Lancet, 249*(6453), 583–586. https://doi.org/10.1016/S0140-6736(47)91856-4

Polanyi, M. (1962). The Republic of Science: Its Political and Economic Theory. *Minerva, 1*(1), 54–73.

Popper, K. (1935). *Logik der Forschung*. Springer.

Popper, K. (1959). *The Logic of Scientific Discovery*. Basic Books.

Poskett, J. (2022). *Horizons: The Global Origins of Modern Science* (First Mariner Books edition). Mariner Books.

Prinz, F., Schlange, T., & Asadullah, K. (2011). Believe it or not: How much can we rely on published data on potential drug targets? *Nature Reviews Drug Discovery, 10*(9), Article 9. https://doi.org/10.1038/nrd3439-c1

Puputti, H. (2022, April 19). *Positioning Social Sciences and Humanities Actors in the Ninth European Union Framework Programme: A Discourse Analysis of Horizon Europe Reference Documents* [Pro gradu -työ]. Laturi.Oulu.Fi. https://oulurepo.oulu.fi/handle/10024/19638

Quarati, A., & Raffaghelli, J. E. (2022). Do researchers use open research data? Exploring the relationships between usage trends and metadata quality across scientific disciplines from the Figshare case. *Journal of Information Science, 48*(4), 423–448. https://doi.org/10.1177/0165551520961048

Radder, H. (2019). *From Commodification to the Common Good: Reconstructing Science, Technology, and Society*. University of Pittsburgh Press.

Rai, A. K., & Sampat, B. N. (2012). Accountability in patenting of federally funded research. *Nature Biotechnology, 30*(10), Article 10. https://doi.org/10.1038/nbt.2382

Ramiro, S., Page, M. J., Whittle, S. L., Huang, H., Verhagen, A. P., Beaton, D. E., Richards, P., Voshaar, M., Shea, B., Windt, D. A. van der, Kopkow, C., Lenza, M., Jain, N. B., Richards, B., Hill, C., Gill, T. K., Koes, B., Foster, N. E., Conaghan, P. G., ... Buchbinder, R. (2019). The OMERACT Core Domain Set for Clinical Trials of Shoulder Disorders. *The Journal of Rheumatology, 46*(8), 969–975. https://doi.org/10.3899/jrheum.181070

Rehn, A. (2019). *Innovation for the Fatigued: How to Build a Culture of Deep Creativity*. Kogan Page Limited.

Rice, D. B., Raffoul, H., Ioannidis, J. P. A., & Moher, D. (2020). Academic criteria for promotion and tenure in biomedical sciences faculties: Cross sectional analysis of international sample of universities. *BMJ, 369*, m2081. https://doi.org/10.1136/bmj.m2081

Rogers, E. M. (1976). Where are we in understanding the diffusion of innovations. In *Communication and Change: The Last Ten Years—And the Next* (pp. 204–222). University Press of Hawaii.

Rolin, K. (2019). The Epistemic Significance of Diversity. In *The Routledge Handbook of Social Epistemology*. Routledge.

Romero, F. (2019). Philosophy of science and the replicability crisis. *Philosophy Compass, 14*(11), e12633. https://doi.org/10.1111/phc3.12633

Rosen, C., Guenther, L., & Froehlich, K. (2016). The Question of Newsworthiness: A Cross-Comparison Among Science Journalists' Selection Criteria in Argentina, France, and Germany. *Science Communication, 38*(3), 328–355. https://doi.org/10.1177/1075547016645585

Rosling, H. (2018). *Factfulness: Ten Reasons We're Wrong About the World—And Why Things Are Better than you Think*. Flatiron Books.

Sachs, J. D., & Kotlikoff, L. J. (2012). *Smart Machines and Long-Term Misery* (Working Paper No. 18629). National Bureau of Economic Research. https://doi.org/10.3386/w18629

Saint-Paul, G. (2008). *Innovation and Inequality: How Does Technical Progress Affect Workers?* Princeton University Press.

Sanberg, P. R., Gharib, M., Harker, P. T., Kaler, E. W., Marchase, R. B., Sands, T. D., Arshadi, N., & Sarkar, S. (2014). Changing the academic culture: Valuing patents and commercialization toward tenure and career advancement. *Proceedings of the National Academy of Sciences, 111*(18), 6542–6547. https://doi.org/10.1073/pnas.1404094111

Santos-Hermosa, G., Quarati, A., Loría-Soriano, E., & Raffaghelli, J. E. (2023). Why Does Open Data Get Underused? A Focus on the Role of (Open) Data Literacy. In J. E. Raffaghelli & A. Sangrà (Eds), *Data Cultures in Higher Education: Emergent Practices and the Challenge Ahead* (pp. 145–177). Springer International Publishing. https://doi.org/10.1007/978-3-031-24193-2_6

Sarewitz, D. (2016). Saving Science. *The New Atlantis*. https://www.thenewatlantis.com/publications/saving-science

Sarstedt, M., Adler, S. J., Ringle, C. M., Cho, G., Diamantopoulos, A., Hwang, H., & Liengaard, B. D. (2024). Same model, same data, but different outcomes: Evaluating the impact of method choices in structural equation modeling. *Journal of Product Innovation Management, 41*(6), 1100–1117. https://doi.org/10.1111/jpim.12738

Schimel, D., & Keller, M. (2015). Big questions, big science: Meeting the challenges of global ecology. *Oecologia, 177*(4), 925–934. https://doi.org/10.1007/s00442-015-3236-3

Schönbrodt, F. D., Wagenmakers, E.-J., Zehetleitner, M., & Perugini, M. (2017). Sequential hypothesis testing with Bayes factors: Efficiently testing mean differences. *Psychological Methods, 22*(2), 322–339. https://doi.org/10.1037/met0000061

Schumpeter, J. A. (1934). *The Theory of Economic Development: An Inquiry into Profits, Capital, Credit, Interest and the Business Cycle*. Transaction Publishers.

Schweinsberg, M., Feldman, M., Staub, N., van den Akker, O. R., van Aert, R. C. M., van Assen, M. A. L. M., Liu, Y., Althoff, T., Heer, J., Kale, A., Mohamed, Z., Amireh, H., Venkatesh Prasad, V., Bernstein, A., Robinson, E., Snellman, K., Amy Sommer, S., Otner, S. M. G., Robinson, D., ... Luis Uhlmann, E. (2021). Same data, different conclusions: Radical dispersion in empirical results when independent analysts operationalize and test the same hypothesis. *Organizational Behavior and Human Decision Processes, 165*, 228–249. https://doi.org/10.1016/j.obhdp.2021.02.003

Silan, M., Adetula, A., Basnight-Brown, D. M., Forscher, P. S., Dutra, N., & IJzerman, and H. (2021). Psychological Science Needs the Entire Globe, Part 2. *APS Observer, 34*. https://www.psychologicalscience.org/observer/psychological-science-needs-the-entire-globe-part-2

Silberzahn, R., Uhlmann, E. L., Martin, D. P., Anselmi, P., Aust, F., Awtrey, E., Bahník, Š., Bai, F., Bannard, C., Bonnier, E., Carlsson, R., Cheung, F., Christensen, G., Clay, R., Craig, M. A., Dalla Rosa, A., Dam, L., Evans, M. H., Flores Cervantes, I., ... Nosek, B. A. (2018). Many Analysts, One Data Set: Making Transparent How Variations in Analytic Choices Affect Results. *Advances in Methods and Practices in Psychological Science, 1*(3), 337–356. https://doi.org/10.1177/2515245917747646

Simmons, J. P., Nelson, L. D., & Simonsohn, U. (2011). False-Positive Psychology: Undisclosed Flexibility in Data Collection and Analysis Allows Presenting Anything as Significant. *Psychological Science*, *22*(11), 1359–1366. https://doi.org/10.1177/0956797611417632

Solovey, M. (2015). *Shaky Foundations: The Politics-Patronage-Social Science Nexus in Cold War America*. Rutgers University Press.

Solow, R. M. (1957). Technical Change and the Aggregate Production Function. *The Review of Economics and Statistics*, *39*(3), 312–320. https://doi.org/10.2307/1926047

Springer Nature (July 2025). The state of null results: Insights from 11,000 researchers on negative or inconclusive results. https://stories.springernature.com/the-state-of-null-results-white-paper/index.html

Stagge, J. H., Rosenberg, D. E., Abdallah, A. M., Akbar, H., Attallah, N. A., & James, R. (2019). Assessing data availability and research reproducibility in hydrology and water resources. *Scientific Data*, *6*(1), 190030. https://doi.org/10.1038/sdata.2019.30

Stamm, J. (2019). Chapter 19: Interdisciplinarity put to test: science policy rhetoric vs scientific practice - the case of integrating the social sciences and humanities in Horizon 2020. In D. Simon, S. Kuhlmann, J. Stamm, & W. Canzler (Eds), *Handbook on Science and Public Policy*. https://www.elgaronline.com/edcollchap/edcoll/9781784715939/9781784715939.00030.xml

Stengers, I. (2018). *Another Science is Possible: A Manifesto for Slow Science* (English edition). Polity Press.

Sterckx, S. (2010). Knowledge transfer from academia to industry through patenting and licensing: Rhetoric and reality. In H. Radder (Ed.), *The Commodification of Academic Research: Science and the Modern University* (Vol. 1–1 online resource (vii, 350 pages): illustrations). University of Pittsburgh Press.

Stevens, J. R. (2017). Replicability and Reproducibility in Comparative Psychology. *Frontiers in Psychology*, *8*. https://doi.org/10.3389/fpsyg.2017.00862

Stiglitz, J. E. (2019). Is Growth Passé? *Project Syndicate*. https://www.project-syndicate.org/commentary/climate-change-demands-transition-to-green-growth-by-joseph-e-stiglitz-2019-12

Stiglitz, J. E., Sen, A., & Fitoussi, J.-P. (2010). *Mismeasuring Our Lives: Why GDP Doesn't Add Up : The Report*. New Press: Distributed by Perseus Distribution.

Straiton, J. (2024). Artificial intelligence: Help or hindrance in solving the reproducibility crisis? *BioTechniques*, *76*(7), 291–294. https://doi.org/10.1080/07366205.2024.2355776

Stroebe, W., & Strack, F. (2014). The Alleged Crisis and the Illusion of Exact Replication. *Perspectives on Psychological Science*, *9*(1), 59–71. https://doi.org/10.1177/1745691613514450

Sundaram, S. S., & Musen, M. A. (2023). *Making Metadata More FAIR Using Large Language Models*. https://doi.org/10.4126/FRL01-006444995

Susskind, D. (2021). *A World Without Work: Technology, Automation and How We Should Respond* (Paperback edition). Penguin Books.

Taleb, N. N. (2012). *Antifragile: Things that Gain from Disorder* (Unabridged). Random House Audio.

Tallon, D., Chard, J., & Dieppe, P. (2000). Relation between agendas of the research community and the research consumer. *The Lancet, 355*(9220), 2037–2040. https://doi.org/10.1016/S0140-6736(00)02351-5

Tebes, J. K., & Thai, N. D. (2018). Interdisciplinary team science and the public: Steps toward a participatory team science. *American Psychologist, 73*(4), 549–562. https://doi.org/10.1037/amp0000281

Teplitskiy, M., Peng, H., Blasco, A., & Lakhani, K. R. (2022). Is novel research worth doing? Evidence from peer review at 49 journals. *Proceedings of the National Academy of Sciences, 119*(47), e2118046119. https://doi.org/10.1073/pnas.2118046119

Thorndike, L. (1951). Newness and Craving for Novelty in Seventeenth-Century Science and Medicine. *Journal of the History of Ideas, 12*(4), 584–598. https://doi.org/10.2307/2707487

Torge, W. (2012). 150 Years of International Cooperation in Geodesy: Precursors and the Development of Baeyer's Project to a Scientific Organisation. *ZFV, 3*, 166–175.

Trafimow, D., Hyman, M. R., & Kostyk, A. (2020). The (im)precision of scholarly consumer behavior research. *Journal of Business Research, 114*, 93–101. https://doi.org/10.1016/j.jbusres.2020.04.008

Trampush, J. W., Yang, M. L. Z., Yu, J., Knowles, E., Davies, G., Liewald, D. C., Starr, J. M., Djurovic, S., Melle, I., Sundet, K., Christoforou, A., Reinvang, I., DeRosse, P., Lundervold, A. J., Steen, V. M., Espeseth, T., Räikkönen, K., Widen, E., Palotie, A., ... Lencz, T. (2017). GWAS meta-analysis reveals novel loci and genetic correlates for general cognitive function: A report from the COGENT consortium. *Molecular Psychiatry, 22*(3), Article 3. https://doi.org/10.1038/mp.2016.244

Tseng, A. A., & Raudensky, M. (2014). Assessments of technology transfer activities of US universities and associated impact of Bayh–Dole Act. *Scientometrics, 101*(3), 1851–1869. https://doi.org/10.1007/s11192-014-1404-6

Tugwell, P., Boers, M., Brooks, P., Simon, L., Strand, V., & Idzerda, L. (2007). OMERACT: An international initiative to improve outcome measurement in rheumatology. *Trials, 8*(1), 38. https://doi.org/10.1186/1745-6215-8-38

UNDP. (2018). *Spark, Scale, Sustain.*

Vaesen, K. (2021). French Neopositivism and the Logic, Psychology, and Sociology of Scientific Discovery. *HOPOS: The Journal of the International*

Society for the History of Philosophy of Science, 11(1), 183–200. https://doi.org/10.1086/712934

Vaesen, K., & Katzav, J. (2017). How much would each researcher receive if competitive government research funding were distributed equally among researchers? *PLOS ONE, 12*(9), e0183967. https://doi.org/10.1371/journal.pone.0183967

Vaesen, K., & Katzav, J. (2019). The National Science Foundation and philosophy of science's withdrawal from social concerns. *Studies in History and Philosophy of Science Part A, 78*, 73–82. https://doi.org/10.1016/j.shpsa.2019.01.001

Valdez, J., Sharma, S., & Gould, J. (2024). *The Shadow of the Ring: How the Gollum Effect Impacts Research and Careers* (SSRN Scholarly Paper No. 4888507). Social Science Research Network. https://doi.org/10.2139/ssrn.4888507

Van Bouwel, J. (2015). Towards Democratic Models of Science: Exploring the Case of Scientific Pluralism. *Perspectives on Science, 23*(2), 149–172. https://doi.org/10.1162/POSC_a_00165

Veugelers, R., Wang, J., & Stephan, P. (2022a). Do Funding Agencies Select and Enable Risky Research: Evidence from ERC Using Novelty as a Proxy of Risk Taking. *NBER Working Papers, 30320*, 1–63.

Veugelers, R., Wang, J., & Stephan, P. (2022b). *Do Funding Agencies Select and Enable Risky Research: Evidence from ERC Using Novelty as a Proxy of Risk Taking*. w30320. https://doi.org/10.3386/w30320

Vinkers, C. H., Tijdink, J. K., & Otte, W. M. (2015). Use of positive and negative words in scientific PubMed abstracts between 1974 and 2014: Retrospective analysis. *BMJ, 351*, h6467. https://doi.org/10.1136/bmj.h6467

Vinsel, L., & Russell, A. L. (2020). *The Innovation Delusion: How Our Obsession with the New has Disrupted the Work that Matters Most*. Currency.

Wagenmakers, E.-J., Wetzels, R., Borsboom, D., & van der Maas, H. L. J. (2011). Why psychologists must change the way they analyze their data: The case of psi: Comment on Bem (2011). *Journal of Personality and Social Psychology, 100*(3), 426–432. https://doi.org/10.1037/a0022790

Waithira, N., Mukaka, M., Kestelyn, E., Chotthanawathit, K., Phuong, D. N. T., Thanh, H. N., Osterrieder, A., Lang, T., & Cheah, P. Y. (2024). Data sharing and reuse in clinical research: Are we there yet? A cross-sectional study on progress, challenges and opportunities in LMICs. *PLOS Global Public Health, 4*(11), e0003392. https://doi.org/10.1371/journal.pgph.0003392

Wallerstein, N., Duran, B., Oetzel, J. G., & Minkler, M. (2018). *Community-Based Participatory Research for Health: Advancing Social and Health Equity* (Third edition). Jossey-Bass, a Wiley Brand.

Wang, H., Fu, T., Du, Y., Gao, W., Huang, K., Liu, Z., Chandak, P., Liu, S., Van Katwyk, P., Deac, A., Anandkumar, A., Bergen, K., Gomes, C. P., Ho, S., Kohli, P., Lasenby, J., Leskovec, J., Liu, T.-Y., Manrai, A., ... Zitnik, M. (2023). Scientific discovery in the age of artificial intelligence. *Nature, 620*(7972),

47–60. https://doi.org/10.1038/s41586-023-06221-2

Wang, K., Goldenberg, A., Dorison, C. A., Miller, J. K., Uusberg, A., Lerner, J. S., Gross, J. J., Agesin, B. B., Bernardo, M., Campos, O., Eudave, L., Grzech, K., Ozery, D. H., Jackson, E. A., Garcia, E. O. L., Drexler, S. M., Jurković, A. P., Rana, K., Wilson, J. P., ... Moshontz, H. (2021). A multi-country test of brief reappraisal interventions on emotions during the COVID-19 pandemic. *Nature Human Behaviour, 5*(8), 1089–1110. https://doi.org/10.1038/s41562-021-01173-x

Webb, R. M., & Kurtz, L. (2022). Politics v. science: How President Trump's war on science impacted public health and environmental regulation. *Progress in Molecular Biology and Translational Science, 188*(1), 65–80. https://doi.org/10.1016/bs.pmbts.2021.11.006

Weinberg, A. M. (1972). Science and Trans-Science. *Minerva, 10*(2), 209–222.

Whitley, R., Gläser, J., & Laudel, G. (2018). The Impact of Changing Funding and Authority Relationships on Scientific Innovations. *Minerva, 56*(1), 109–134. https://doi.org/10.1007/s11024-018-9343-7

Wisnioski, M. H. (2025). *Every American an Innovator: How Innovation Became a Way of Life.* The MIT Press. http://mitpress.mit.edu/9780262550734

Wit, M. P. T. de, Abma, T. A., Loon, M. S. K., Collins, S., & Kirwan, J. (2014). What Has Been the Effect on Trial Outcome Assessments of a Decade of Patient Participation in OMERACT? *The Journal of Rheumatology, 41*(1), 177–184. https://doi.org/10.3899/jrheum.130816

Wu, L., Wang, D., & Evans, J. A. (2019). Large teams develop and small teams disrupt science and technology. *Nature, 566*(7744), 378–382. https://doi.org/10.1038/s41586-019-0941-9

Yang, Y., Youyou, W., & Uzzi, B. (2020). Estimating the deep replicability of scientific findings using human and artificial intelligence. *Proceedings of the National Academy of Sciences, 117*(20), 10762–10768. https://doi.org/10.1073/pnas.1909046117

Yarkoni, T. (2022). The generalizability crisis. *Behavioral and Brain Sciences, 45*, e1. https://doi.org/10.1017/S0140525X20001685

Zeng, A., Shen, Z., Zhou, J., Fan, Y., Di, Z., Wang, Y., Stanley, H. E., & Havlin, S. (2019). Increasing trend of scientists to switch between topics. *Nature Communications, 10*(1), Article 1. https://doi.org/10.1038/s41467-019-11401-8

Zilwa, D. de. (2005). Using Entrepreneurial Activities as a Means of Survival: Investigating the Processes used by Australian Universities to Diversify their Revenue Streams. *Higher Education, 50*(3), 387–411. https://doi.org/10.1007/s10734-004-6359-8

Zuiderwijk, A., Shinde, R., & Jeng, W. (2020). What drives and inhibits researchers to share and use open research data? A systematic literature review to analyze factors influencing open research data adoption. *PLOS ONE, 15*(9), e0239283. https://doi.org/10.1371/journal.pone.0239283

Index

About the Team

Alessandra Tosi was the managing editor for this book.

Annie Hine proof-read and indexed this manuscript.

Jeevanjot Kaur Nagpal designed the cover. The cover was produced in InDesign using the Fontin font.

Annie Hine typeset the book in InDesign. The main text font is Tex Gyre Pagella and the heading font is Noto Serif.

Jeremy Bowman produced the PDF, paperback, and hardback editions and created the EPUB.

The conversion to the HTML edition was performed with epublius, an open-source software which is freely available on our GitHub page at https://github.com/OpenBookPublishers

Lucy Barnes was in charge of marketing.

This book was peer-reviewed by Prof. Thomas Reydon, Institute of Philosophy, Leibniz University, and an anonymous referee. Experts in their field, these readers give their time freely to help ensure the academic rigour of our books. We are grateful for their generous and invaluable contributions.

This book need not end here...

Share

All our books — including the one you have just read — are free to access online so that students, researchers and members of the public who can't afford a printed edition will have access to the same ideas. This title will be accessed online by hundreds of readers each month across the globe: why not share the link so that someone you know is one of them?

This book and additional content is available at
https://doi.org/10.11647/OBP.0507

Donate

Open Book Publishers is an award-winning, scholar-led, not-for-profit press making knowledge freely available one book at a time. We don't charge authors to publish with us: instead, our work is supported by our library members and by donations from people who believe that research shouldn't be locked behind paywalls.

Join the effort to free knowledge by supporting us at
https://www.openbookpublishers.com/support-us

You may also be interested in:

Hanging on to the Edges

Essays on Science, Society and the Academic Life

Daniel Nettle

https://doi.org/10.11647/OBP.0155

The Scientific Revolution Revisited

Mikuláš Teich

https://doi.org/10.11647/OBP.0054

Animals and Medicine

The Contribution of Animal Experiments to the Control of Disease

Jack Botting and Regina Botting (eds); foreword by Adrian R. Morrison

https://doi.org/10.11647/OBP.0055